OPTICAL SENSORS - NEW DEVELOPMENTS AND PRACTICAL APPLICATIONS

Edited by **Mohamad Yasin, Sulaiman Wadi Harun and Hamzah Arof**

Optical Sensors - New Developments and Practical Applications

http://dx.doi.org/10.5772/57077

Edited by Mohamad Yasin, Sulaiman Wadi Harun and Hamzah Arof

Contributors

Ashok K Sood, Hirotaka Sakaue, Marcus Wolff, Henry Bruhns, Nunzio Cennamo, Michal Lucki, Leos Bohac, Richard Zeleny, Fabricio Povh, Wagner dos Anjos, Chengbo Mou, Zhijun Yan, Kaiming Zhou, Lin Zhang, Alsu Akhmetshina, Ali Rostami, Chien-Hung Yeh, Chi-Wai Chow

Published by InTech

Janeza Trdine 9, 51000 Rijeka, Croatia

Publishing Process Manager Iva Lipovic

Technical Editor InTech DTP team

Cover InTech Design team

First published March, 2014

Printed in Croatia

Additional hard copies can be obtained from orders@intechopen.com

Optical Sensors - New Developments and Practical Applications, Edited by Mohamad Yasin, Sulaiman Wadi Harun and Hamzah Arof

p. cm.

ISBN 978-953-51-1233-4

For EU product safety concerns:
IN TECH d.o.o., Prolaz Marije Krucifikse Kozulić 3, 51000 Rijeka, Croatia,
info@intechopen.com or visit our website at intechopen.com.

Meet the editors

Mohamad Yasin received his B.Sc in 1990 from Airlangga University. He went on to receive M.Sc and Ph.D from Gadjah Mada University in 1999 and 2010, respectively. His research interest includes fiber optic sensors and devices. Currently, he is an associate professor at the Faculty of Science and Technology, Airlangga Univesity, Indonesia.

Sulaiman Wadi Harun received the B.E degree in Electrical and Electronics System Engineering from Nagaoka University of Technology, Japan in 1996, and M.Sc and Ph.D degrees in Photonic Technology from University of Malaya in 2001 and 2004, respectively. He is actively working on optical amplifiers, fiber lasers and fiber-optic sensors, with more than 350 publications in ISI journals and citations of more than 2200. He is currently a full Professor in the Department of Electrical Engineering, University of Malaya, Malaysia and his email address is swharun@um.edu.my

Hamzah Arof received his B.Sc from Michigan State University, and Ph.D from the University of Wales, all in Electrical engineering. His current research interests include signal processing and photonics. Currently he is affiliated with the Department of Electrical Engineering, University of Malaya, Malaysia. He can be contacted at ahamzah@um.edu.my.

Contents

Preface

This book contains 10 chapters focusing on the new developments and practical applications of optical sensors. Laboratory experiments and field investigations of new and emerging optical sensors are presented. Some data analysis methods are also described. Chapter 1 discusses principle and practical implementation of fiber-optic and free space Michelson interferometer for sensor applications. Considering the coherence length of a He-Ne laser, the Michelson interferometer potentially allows for measurement of far objects, where a signal is guided in an optical fiber. In addition, the interferometer can be miniaturized or built as a fiber optic based sensor. Chapter 2 reviews recent development of excessively tilted fiber grating as optical sensors for various physical parameters such as strain, twist, loading, refractive index (RI) and liquid level sensing. The use of mesoporous polymers for optical sensors is discussed in Chapter 3. Chapter 4 describes the use of photonic crystal laser for gas sensor that provides some advantages over the ones based on traditional semiconductor lasers. The emission and operation parameters of this laser as well as measured spectra of H2O at 1910 nm are also presented in this chapter. Chapter 5 presents Whispering Gallery Mode (WGM)-based and other fiber optic sensors for biomedical applications.

Fiber Bragg gratings (FBGs) have been widely employed for various optical fiber sensing applications. In Chapter 6, two different FBG-based sensor networks are proposed and demonstrated utilizing passive system design. The first design is based on a 25 km cavity length erbium-doped fiber (EDF) ring laser for detecting multiple FBG sensors in the network. The second design is based on a multi-ring passive sensing architecture, which does not have any active components in the entire network. Thus, the proposed system can facilitate the construction of a reliable FBG sensing network for large-scale and multi-point architecture. Chapter 7 presents the development of bio and chemical sensors based on Surface Plasmon Resonance (SPR) in a plastic optical fiber (POF). The POF sensor configurations are presented to monitor an aqueous environment (refractive index around 1.333), with a resolution ranging from 10-4 to 10-3 (RIU). A wide range of optical sensors have been utilized in agriculture, which include sensors used to analyze soil attributes to the ones installed in combines to measure protein content in wheat grains while they are being harvested. Chapter 8 specifically discusses optical sensors to measure agricultural crop reflectance. Chapter 9 presents recent advances in nanostructured based detector technology, materials and devices for optical sensing applications. An overview of recent works on a variety of semiconductors and advanced materials such as GaN, ZnO, Si/SiGe, InGaAs and CNT for optical sensing applications is thoroughly discussed in this chapter. In the final chapter, a new luminophore application method is proposed and demonstrated based on dipping deposition in the fabrication of an optimized optical pressure sensor for unsteady aerodynamic applications.

We hope this book chapter compilation will provide the readers with a broad overview and sampling of the innovative research in optical sensors.

Mohamad Yasin
Department of Physics,
Faculty of Science and Technology,
Airlangga University, Indonesia

Sulaiman Wadi Harun
Department of Electrical Engineering, University of Malaya,
Malaysia

Hamzah Arof
Department of Electrical Engineering, University of Malaya,
Malaysia

Chapter 1

Fiber Optic and Free Space Michelson Interferometer — Principle and Practice

Michal Lucki, Leos Bohac and Richard Zeleny

Additional information is available at the end of the chapter

http://dx.doi.org/10.5772/57149

1. Introduction

Michelson interferometer is used in metrology of small amplitude nonelectric physical quantities for its accuracy, noncontact and noninvasive procedure. It is broadly used in sensor applications. There are many papers assuming the use of an interferometer and focusing on measured results, but there are not many works offering practical knowledge on how to construct and run Michelson interferometer. In this chapter we discuss wide range of aspects, which greatly facilitate the launch of Michelson interferometer in in-situ conditions.

Random addition to a signal can practically disqualify many techniques for their eventual application in accurate measurements of displacement or vibrations. The interferometric method is suitable for signals that require a noninvasive and noncontact method [1]. It allows avoiding a physical contact with a measured object that would originate spurious signals causing errors greater than the values to be measured. Such signals may be encountered in industrial applications like mining or construction technologies, in measurements of resonant frequencies of machines or bridges and last but not least in the measurement of small deformations or spatial distributions of temperature. Finally, measurements performed in harsh environment, such as the ones with extremely high temperature, could damage the measuring apparatus. In addition, the interferometric method is attractive for its price.

2. Principle of the method

The necessary and sufficient condition to observe interference of light is the coherence length being greater than optical path difference between two superposed beams. In practice, interference is considered for coherent waves, where the maximum distance from the light source, at which interference is observed, is related to the spatial cross-correlation between two points in a wave for an arbitrarily selected time instant.

A very good description of interference observed in Michelson interferometer as well as many practical aspects about running experiment with Michelson setup is presented in [2]. A numerical model of a Michelson interferometer is published in [3]. A block scheme of a Michelson interferometer is presented in Figure 1. Based on the scheme, in the following section, we present the mathematical model of light interference in a free space Michelson interferometer.

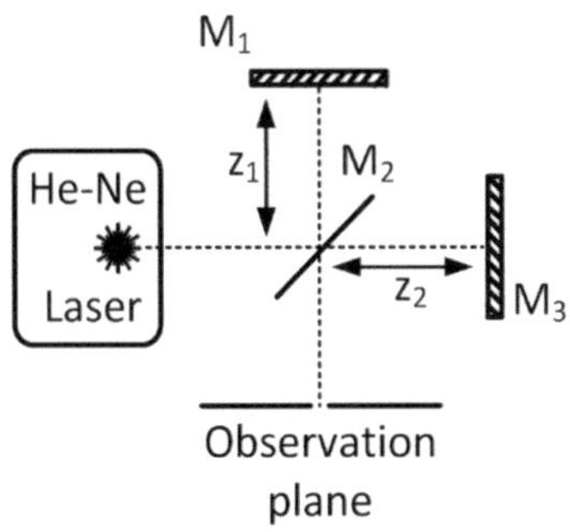

Figure 1. Schematic diagram of a Michelson interferometer. M_2 – 50% mirror (beam splitter), M_1, M_3 – dielectric mirrors.

2.1. Interference of light observed in a free space Michelson interferometer

Let us consider a wave entering the Michelson interferometer, presented in Figure 1.

$$E = E_0 \sin\left(kz - \omega t + \varphi_0\right) \tag{1}$$

Where k is known as angular wavenumber that is to express the magnitude of the wave vector, and for dielectrics is as:

$$k = \frac{2\pi}{\lambda} \tag{2}$$

And ω is angular frequency. Knowing the frequency of light ν having a wavelength λ and a speed c, it can be written as:

$$\omega = 2\pi\nu = 2\pi\frac{2}{\nu} \tag{3}$$

For simplicity and without restricting generality, we assume certain time instant where φ_0 is equal to 0.

$$E = E_0 \sin(kz - \omega t) \tag{4}$$

The wave is split at a beam splitter in Michelson interferometer and two waves E_1 and E_2 are formed. The wave that is partially reflected at the beam splitter M_2 and is reflected at M_1 mirror can be written as:

$$E_1 = E_{01} \sin(kz_1 - \omega t) \tag{5}$$

The wave transmitted through the beam splitter and reflected from M_3 mirror is as:

$$E_2 = E_{02} \sin(kz_2 - \omega t) \tag{6}$$

where z_1 is optical path length from the beam splitter (M_2) to M_1 mirror, z_2 is optical path length from the beam splitter (M_2) to M_3 mirror, as displayed in Figure 1.

E_0 amplitude is split to E_{01} and E_{02}. In theory, both waves do not have any phase shift with regard to each other as well as in an ideal case the splitter should separate them symmetrically. In real Michelson setup, splitting is not symmetrical and the amplitudes of split waves are:

$$E_{01} = E_0 \cdot r_2 \cdot r_1 \tag{7}$$

$$E_{02} = E_0 \cdot t_2 \cdot r_3 \tag{8}$$

Where r_2 is the reflection coefficient of a beam splitter, r_1 is the reflection coefficient of M_1 mirror, t_2 is the transmission coefficient of a beam splitter, and r_3 is the reflection coefficient of M_3 mirror. In addition, in practical consideration, the phase of E_2 wave transmitted through the beam splitter is changed, as referred to the phase of E_1. Then, they should be denoted as:

$$E_1 = E_{01} \sin(kz_1 - \omega t) \tag{9}$$

$$E_2 = E_{02} \sin(kz_2 - \omega t + \varphi) \tag{10}$$

The superposition of interfering waves at the beam splitter is the sum of both waves, denoted as E_3:

$$E_3 = E_1 + E_2 = E_{01}\sin(kz_1 + \omega t) + E_{02}\sin(kz_2 + \omega t + \varphi) \tag{11}$$

A photodetector detects light intensity I related to intensity of electric field E, as in (12):

$$I \sim E^2 \tag{12}$$

For I_3 as the intensity related to E_3 it can be written:

$$I_3 \sim \left[E_{01}\sin(kz_1 + \omega t) + E_{02}\sin(kz_2 + \omega t + \varphi)\right]^2 \tag{13}$$

$$\begin{aligned} I_3 \sim & E_{01}^2\sin^2(kz_1 + \omega t) + 2E_{01}E_{02}\sin(kz_1 + \omega t)\sin(kz_2 + \omega t + \varphi) + \\ & + E_{02}^2\sin^2(kz_2 + \omega t + \varphi) \end{aligned} \tag{14}$$

To simplify, the following trigonometric operation can be performed on second term:

$$2\sin\alpha \cdot \sin\beta = \cos(\alpha - \beta) + \cos(\alpha + \beta) \tag{15}$$

We obtain:

$$\begin{aligned} I_3 \sim & E_{01}^2\sin^2(kz_1 + \omega t) + E_{01}E_{02}\cos\left[k(z_1 - z_2) - \varphi\right] + \\ & + E_{01}E_{02}\cos\left[k(z_1 + z_2) + 2\omega t + \varphi\right] + E_{02}^2\sin^2(kz_2 + \omega t + \varphi) \end{aligned} \tag{16}$$

It can be noticed that second term $E_{01}E_{02}\cos\left[k(z_1 - z_2) - \varphi\right]$ is time independent, first and fourth terms oscillate at ω, third term oscillates at 2ω. Using this value of λ and the speed of light being c=3x10^8 m.s^{-1}, the frequency of light ν is calculated as in (3). Since the *sin* term of the first and last term oscillates at this frequency, the *cos* term oscillates at double of this frequency, the frequency changes are too fast for a detector to react (it is few ranges slower) [2]. It instead measures average value of these terms. For *sin*^{2} term it is 0.5 and for *cos* it is 0. Eq. (16) can be rewritten to:

$$I \sim \frac{1}{2}E_{01}^2 + \frac{1}{2}E_{02}^2 + E_{01}E_{02}\cos(k\Delta z - \varphi) \tag{17}$$

where Δz is optical path difference:

$$\Delta z = z_1 - z_2 \tag{18}$$

http://dx.doi.org/10.5772/57149

Intensity I is maximum if and only if *cos* term is equal to 1, i.e. when $k\Delta z$-φ is zero or multiple of 2π. Minimum intensity is if and only if *cos* term is equal to -1. Then, it can be written:

$$I_{\max} \sim \frac{1}{2}E_{01}^2 + \frac{1}{2}E_{02}^2 + E_{01}E_{02} = \frac{1}{2}\left(E_{01} + E_{02}\right)^2 \tag{19}$$

$$I_{\min} \sim \frac{1}{2}E_{01}^2 + \frac{1}{2}E_{02}^2 - E_{01}E_{02} = \frac{1}{2}\left(E_{01} - E_{02}\right)^2 \tag{20}$$

Light intensity depends on the optical path difference Δz. If the optical path difference of both interferometer's arms is equal or is multiple of 2π, there is constructive interference and the intensity of interference fringes is maximum.

The above model refers to a free space Michelson interferometer. The mathematical apparatus for a fiber optic Michelson interferometer can be more extensive, since one could consider the reflections at connectors, fiber dispersion, as well as the properties of polarizers, a coupler and the optional presence of a fiber stretcher. However, the model reflects the general idea on light interference in any Michelson interferometer.

3. Experimental setup

When planning how to arrange an experiment using an interferometer, the choice can be made between large number of interferometers, as for example Michelson interferometer [4], Mach-Zehnder interferometer [5], Sagnac interferometer [6], Fabry Perot interferometer, and last but not least Fizeau interferometer, Twyman-Green interferometer, Nomarski interferometer [7]. The selection of interferometer is not always arbitrary; it must be suited to the application. In addition, it is possible to build up an interferometer with a free space or a fiber optic arrangement. The Michelson and Mach-Zehnder are the most widely used interferometers. Michelson interferometer is used for its simplicity, accuracy, and lower number of components. In this paper we focus on the Michelson interferometer, in which both interference beams propagate in their arms twice on the contrary to Mach-Zehnder interferometer, which is simpler and easier to implement. The Michelson interferometer also provides easier calibration and better control of mechanical stability.

Obtaining interference fringes is just one of many steps in the measurement procedure. Another step is the integration of a measured object with the interferometer. Transparent objects can be inserted into a beam path in one arm. Other objects should be placed at the end of the interferometer arm or fixed to the mirror.

There is a number of aspects fundamental for experiment like thermal stability of light source used, as well as its coherence length. The experimental setup must suppress ambient vibrations responsible for the spurious signal and thus it must be mechanically stable; otherwise the interference pattern is not stable in time. Especially in free space arrangement, where mechanical stability of the setup is required, the use of optical table suppressing vibrations is practically unavoidable. Fiber optic arrangement can suppress disturbing signals, since signal is guided in optical fibers, whose length is barely influenced by mechanical instability.

An interferometer should be mounted on a pneumatic laboratory table, for example in the faculty cellars where vibrations of ground are minimal, and mounted by using special holders. In very accurate measurements, where measured quantity is a multiple of the wavelength, it is necessary to use lasers with good temperature stability of a wavelength.

3.1. Free space Michelson interferometer

In free space Michelson interferometer, it is suitable to use a laser from the visible range to facilitate calibration and data reading, for example a He-Ne laser, which exhibits good wavelength stability (approximately 2 MHz). In free space Michelson interferometer, the position of mirrors and a beam splitter is adjusted by observing the incident light in the interference plane. Although this arrangement is very sensitive to mechanical instability, the fringes can be obtained employing a systematic approach. To calibrate the interferometer, the angles between mirrors are adjusted as well as proper angles between the beam splitter and mirrors are set to get the interference pattern with required number of fringes.

Adjusting the position of a beam splitter affects both beams at once and is more time consuming. The position of both dielectric mirrors should be set, and once both beams overlap at the beam splitter, the position of all elements of the setup is adjusted to set right distribution of fringes in the observation plane. The increase in arm's length by the distance equal to the value of the half of the wavelength (316.4 nm) causes the change to the phase difference between two beams equal to 2π (one interference fringe). The interferometer is then able to measure displacement in a range of several hundred nanometers (corresponding to the ability to distinguish one fringe) [8].

The concept of a free space Michelson interferometer with a sample investigated object and a connected detector is displayed in Figure 2. A He-Ne laser is one of very broadly used in free space implementations because it emits light from the visible range and exhibits great wavelength stability and decent output power. A laser beam is split into two beams by a beam splitter (M_2), (amplitude splitting does not affect polarization). Each beam is propagated along its optical path controlled by a mirror system (M_1, M_3). The reference beam is guided to the observation plane going twice through the beam splitter. The measuring beam is guided to the investigated object, a copper rod terminated with a dielectric reflecting mirror. (It can potentially be any other object specific for its length, and being solid in the investigated range of temperatures). Measured signal affects the position of the M_3 mirror placed at its end and causes changes to the reflected optical path length in a measuring arm.

The reference and measuring beams interfere at the beam splitter. An additional phase shift between both arms causes changes to the distribution of interference fringes that are counted and monitored. Information about the amount of fringes is stored in a computer and is used to calculate the change in optical path length. Number of fringes passing through the aperture

http://dx.doi.org/10.5772/57149

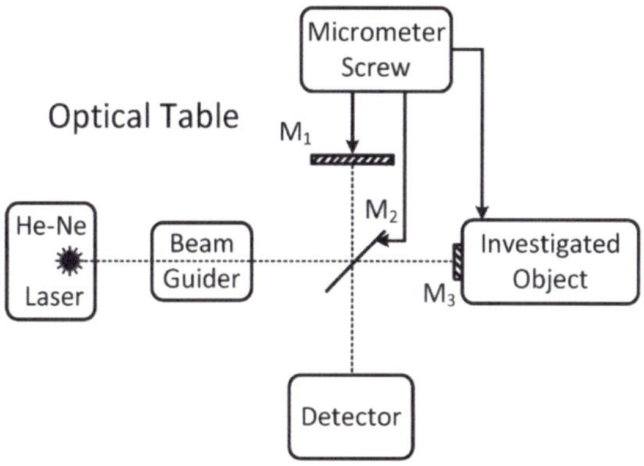

Figure 2. Schematic diagram of a free-space Michelson interferometer. M_2 – 50% mirror (beam splitter), M_1, M_3 – dielectric mirrors.

in an observation plane is proportional to the measured quantity. Sample interferograms obtained in an exemplary experiment are shown in Figure 3.

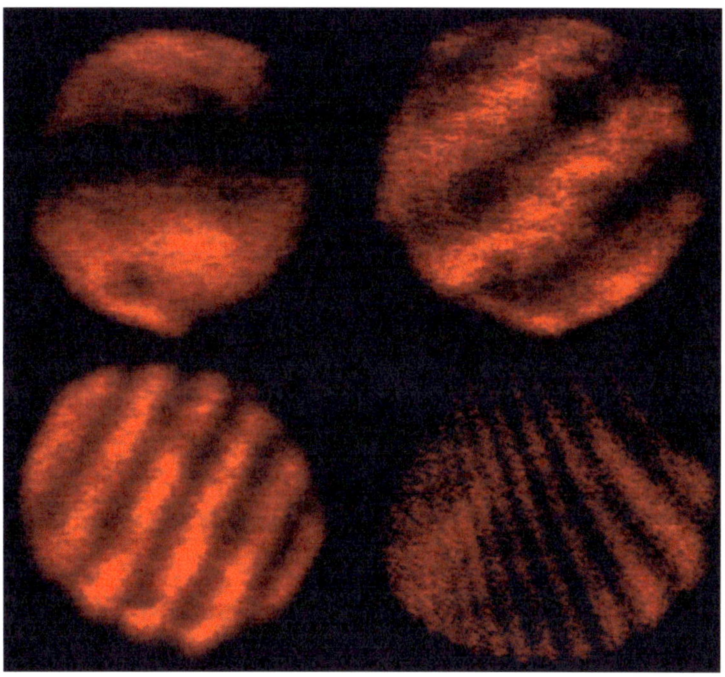

Figure 3. Sample distribution of interference fringes. Mirrors adjusted to one dark fringe (top left), two light and two dark fringes (top right), new fringes appearing when taking measurement (bottom left), changed fringe direction caused by different angle between M_1 and M_3 mirrors (bottom right).

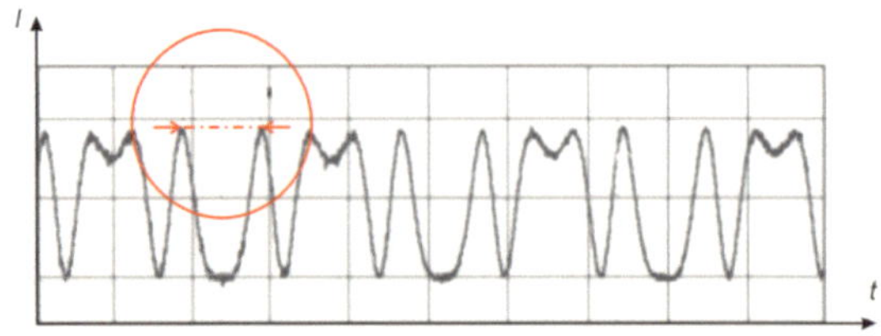

Figure 4. Sample time variation of light intensity at the output of an interferometer measured and displayed on an oscilloscope. Two marked maximum points show one fringe period corresponding to the phase shift of 2π, which means the arm length is changed by the length of the propagated wave.

Considering the fact that fringes pass through the surface of a photodetector, a signal detected by a photodiode or eventually displayed on an oscilloscope contains many minima and maxima of light intensity, such as one displayed in Figure 4. This may be the signal you see after launching the measurement, indicating that the setup started to work. If the interferogram is stable in time and the detected light intensity is a constant value, it means that the investigated object generates no signal or there is no interference for any reason (for example there is a problem with the coherence length of used light source or a detector detects light from one arm only), which in practice can be very confusing. The use of visible light facilitates to verify the existence of interference fringes. In free space interferometers, performing the measurement in the visible spectrum is appropriate because of exact calibration based on a visual control. In fiber optic interferometer, the use of visible light is rather impossible because most of optical fibers are transparent for longer wavelengths.

3.2. Fiber optic Michelson interferometer

Fiber optic Michelson interferometer employs the same principle of splitting a laser beam and inserting the optical path difference between the arms. Both waves interfere at a coupler. However, there are many features specific for fiber optic interferometers, disregarding the fact that we deal with the Michelson interferometer. Essential modifications result from the fact that light is guided in optical fibers. The concept of a fiber optic Michelson interferometer with connected detector is displayed in Figure 5.

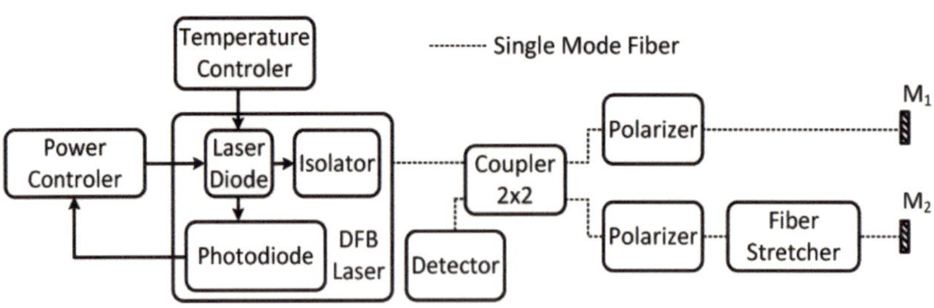

Figure 5. Schematic diagram of a fiber-optic Michelson interferometer. M_1, M_2 – dielectric mirrors.

Fiber mirrors and a coupler should be compatible with the used laser. In theory, they are designed to work at a specific wavelength, but in practice, cheap fiber mirrors are created by using dielectric material and they are broadband. Pigtailed fiber mirrors are commercially available. For example, if you have a mirror labeled as operating at 1310 nm, it is likely that it exhibits high reflectance also at 1550 nm. However, in case of fiber couplers, the splitting ratio, important for fringe thickness and contrast, can be very different at both wavelengths.

Calibration of a fiber optic interferometer requires using additional components, compared to a free space interferometer, where optical path length can be simply adjusted by positioning of mirrors. In fiber optic Michelson interferometer the path length can be increased by inserting an air gap between two fibers, which results in additional losses at the aperture and abandoning the ideal of all-fiber interferometer. One of the most spread techniques is to use a fiber stretcher. A silica fiber is elastic and can be mechanically stretched by about 2 % [9]. It allows for changing the optical path length necessary for the exactly out of phase beams and the near-zero light intensity at a detector. The silica fiber can be stretched by using a cylinder made of a rubber material placed between retainers, which are precisely tightened to each other by using a micrometer screw. Consequently, the cylinder expands and the fiber is stretched, however with the sub-millimeter precision insufficient for the phase adjustment of the light wave. Thus, for a sub-micrometer stretch, a piezoelectric actuator [10] with the voltage control is optimal. Electric field applied to the actuator below its Curie point causes the ions shifts, (an elastic strain), resulting in linear increase (or decrease) in length of the actuator according to the electric field polarity. This feature refers to crystals with no center of symmetry, such as quartz, lithium niobate or bismuth titanate. Most of commercially available fiber stretchers provide the optical path length variation of few micrometers per volt. The fiber stretch is frequency dependent, which should be taken into account, when alternating current is applied to an actuator.

Polarizers in a fiber optic Michelson interferometer are necessary to avoid random polarization due to stretches or birefringence resulting in intensity instability. It can improve the stabilization of visibility in fiber optic sensors [11]. Instead of polarizers that are source of insertion losses, it is suitable to use polarization maintaining fibers.

3.3. Lasers for interferometers

The fiber optic Michelson interferometer require the use of lasers operating at telecommunication wavelengths since optical fibers transmit optical radiation in the range of about 800 – 1600 nm. Helium-Neonium laser, very popular in free space optics, cannot be used with most of commercially available fibers. Distributed Feedback Lasers (DFB) operating at any wavelength from this range are suitable; the most frequently used DFBs work at 1310 nm or at 1550 nm. DFB lasers are cheap, offer relatively narrow spectrum of emitted wavelengths; however, the stability of wavelength is usually worse than in He-Ne lasers. One has to consider the fact that detected number of fringes detected at the observation plane depends on wavelength. Because of this fact, stabilized lasers with a temperature controller can be used. This makes the arrangement more expensive. Since DFB lasers usually work stable in temperature a bit lower than room temperature, one has to wait few minutes until the temperature of a chip is achieved and the emitted wavelength is stabilized. In addition, since the emitted wavelength is not from the visible range, the feedback about the interference fringes is not available for human eye. It is necessary to use a detection circuit (e.g. a photodiode and an oscilloscope) to observe time changes of an interferogram at an observation point. In general, low fringe contrast indicates that some adjustments are required.

An interferometer with long arms must fulfill the condition of coherence length (simplifying, the path taken of all the interfering waves must differ by less than a coherence length, which differs for a LED, white light and lasers). Otherwise interference is not observed. The best coherence length refers to lasers. For example the coherence length of a single mode He-Ne laser can exceed 100 m, while some LED sources exhibit the coherence length of few centimeters.

Objects absorbing the laser beam's energy require using stronger optical power. For this purpose, one can use Argon ion lasers operating at 510 or 490 or 350 nm, exhibiting the power reaching few Watts [7]. Laser diodes, on the contrary, can offer the variety of available wavelengths from infrared to ultraviolet region, however, usually not exceeding the power of tens mW. Multimode lasers usually have worse coherence length than single mode lasers. CO_2 lasers operating in the infrared region can be useful for measurements over long distances.

Wavelength stability influences the accuracy of results. The change in length of the measuring arm of an interferometer is the multiple of wavelength, as expressed in the formula (21).

$$\Delta l = 0.5 \cdot \lambda \cdot n \tag{21}$$

where λ is wavelength, n is number of counted fringes. He-Ne laser exhibits good wavelength stability, for stabilized lasers is fraction of MHz, non-stabilized lasers are usually one range of order worse. Pay attention to older lasers, the real operation wavelength range can differ from nominal range. The knowledge on wavelength instability is often required to evaluate the accuracy of results dependent on λ. If you do not know the wavelength tolerance, you can calculate it from the formula (22):

$$\Delta \lambda = \frac{\lambda_c^2}{c} \Delta f \tag{22}$$

For example, for a He-Ne laser operating at λ=632.8 nm ±2 MHz (as usually expressed in a catalogue), the equivalent expression is λ=632.8 nm ± 2.7 10^{-6} nm, from which it can be concluded that the wavelength instability is negligible.

$$\Delta \lambda = \frac{\lambda_c^2}{c} \Delta f = \frac{\left(632.8 \cdot 10^{-9}\right)^2}{3 \cdot 10^8} \cdot 2 \cdot 10^6 = 2.7 \cdot 10^{-15} m = 2.7 \cdot 10^{-6} nm \tag{23}$$

A practical remark referring to lasers used in interferometry refer to a safety class of used laser. It is recommended to check the safety class, which should be displayed on a label and informs about the necessary protection, which can for example be the use of special protection glasses. In addition, it is not recommend looking into the laser's aperture and avoiding the eye contact with the back reflections from any surface. The direct beam from even a low-power He-Ne laser (0.5 mW) can cause serious eye damage [7].

3.4. Detection of fringes and evaluation of measurement

Different detection methods can be used based on the properties of the measured signals (high speed or slow signals, periodic or nondeterministic signals, low amplitude signals) as well as on the properties of used laser. In general, detection of fringes allows using different devices, such as photodiodes, photomultipliers, photoconductive detectors, CCD sensors or pyroelectric detectors. Photodiodes as the most popular and universal detectors cover wide range of wavelengths, photoconductive detectors are used in the infrared region. Pyroelectric detectors are sensitive over the infrared region, but in addition, they respond to changes in illumination. However, the most sophisticated detection devices assume a detection unit integrated with the circuits for signal processing, including amplification, Fourier transform of a signal suitable for monitoring the frequency changes of an investigated process, and last but not least, converting the measured signal for its transmission (sampling, quantization, encoding, and many others). One of the cheapest solutions is to use commercially available fringe counters or a photodiode with an operational amplifier and to process the signal on a computer. For some applications (i.e. those assuming slow time variations of fringe distribution with the frequency much lower than 1 Hz), it is sufficient to use a spectral analyzer.

Most of affordable optical fibers as well as the majority of fiber optic components are designed to operate at 1300 or 1550 nm and exhibit huge attenuation in the visible range.

When the optical path length of one arm is changed, the distribution of interference fringes on an observation point varies in time. It is possible to monitor their shift, direction and density. For some applications in low speed signal measurement, a spectral analyzer is sufficient, but for signals requiring fast sampling, huge amplification and low noises, a photodiode with an operational amplifier and an oscilloscope is optimal. Another solution is the use of fringe counters; however, they offer poor feedback compared to specialty purposes photodetectors.

The emergence of new fringes is relatively slow from the perspective of a detector. Number of samples measured by a detector per unit of time must be at least twice of the maximum number of occurring interference fringes per unit of time, as according to the Nyquist theorem [12]. For detection of weak signals (few nanowatts), we use a fast response, low noise photodiode integrated with an operational amplifier. Time variation of light intensity is displayed on an oscilloscope.

The measurements can be done with optical shielding to forbid the reception of the disturbing signals. For this purpose, the photodetector is usually bounded into a box with an aperture (without diffraction). To count fringes, the aperture accepts one fringe at a moment. The interferometer is calibrated to display one fringe in the observation plane. More fringes in the aperture could cause that a new fringe is unnoticed.

Localization of fringes in Michelson interferometers can be done using Fourier optics. The analysis presented in [13] helps to understand the Michelson fringes. It can be used for testing optical elements such as mirrors, beam splitters, polarizers and collimation of lenses.

Knowing the number of fringes detected on an observation plane, it is possible to calculate the change in length of the measuring arm from the formula (21). Changes to a measured quantity is proportional to variation in measuring arm length, it is assumed the relation between measured quantity and measuring arm is known. Initial values of the measured quantity must be known since Michelson interferometer is not able to measure absolute values (such as temperature, length) but their differential from the initial values (such as temperature changes, elongation etc.). However, the modified circuits can provide the information about the absolute values, there is a paper about absolute distance meter [14].

A separate problem is the accuracy of obtained results. Repeatability of results could roughly confirm accuracy. In addition, there are few techniques to verify it in theory. The precision of Michelson setup reaches the order of half-wavelength of the He-Ne laser. One fringe corresponds to the change in path length equal to 0.5x632.8 nm, assuming the He-Ne laser. The greater precision can be achieved by using a shorter wavelength. However, much shorter wavelengths are not compatible with most of affordable components. Few methods exist to estimate measurement errors from gathered data and determine resultant accuracy of the calculated results. The most popular one is the exact difference method [15]. The method is suitable for estimation of overall accuracy of multi-tasking measurement and is widely used in physics and engineering, for its statistical accuracy and simplicity. The method can evaluate the optical part of experiment, i.e. Michelson interferometer, but it also enables to combine its accuracy with accuracy of other accompanying measurements, necessary to calculate the result. The general idea is expressed by the formula (24).

$$\Delta a = \sum_{i=1}^{n} \left| \frac{\partial a}{\partial q_i} \right| \cdot \Delta q_i \tag{24}$$

Where a is measured quantity (i.e. parameter, coefficient) being the function of particular quantities directly measured on an investigated object q_i and Δ is inaccuracy). For example, q in formula (24) can be wavelength accuracy or number of read fringes. It can be reading the initial value of measured quantity, the precision of initial value taken from a catalog or accuracy of any accompanying measurement, e.g. time, temperature etc. (The result can be the function of arm's length changes as well as the function of many other processes, not measured by the interferometer). Exact difference method allows combining the contribution of all the measured quantities to overall accuracy.

In practice, results should have the desired form, i.e. they should be stored in a file for further processing on a computer, such as Fourier transform, which is suitable for measuring frequency of vibrations etc. The processing circuit has been proposed in [16].

http://dx.doi.org/10.5772/57149

4. Exemplary experiments

The most investigated application of a Michelson interferometer is a fiber optic sensor [17]. It can measure displacement [18], length changes, amplitude of vibrations (e.g.) the amplitude of seismic waves [19], resonant frequency of vibration, changes in temperature [20], strain (perimeters) [21], and many others. We present the applications to a measurement of thermal expansion coefficient by using a free-space Michelson setup and the application to a measurement of amplitude of vibration of a loudspeaker and its frequency by using a fiber optic Michelson interferometer. Both examples offer a good illustration of the principle of operation of Michelson interferometer and help to understand the procedure of how to build an interferometer.

4.1. Measurements of linear expansion and calculation of thermal expansion coefficient using free space Michelson interferometer

There are few approaches to how to obtain the thermal expansion coefficient; however, high accuracy can be achieved by using noncontact optical methods. A generally accepted view is that if solid is heated, its volume increases. General volumetric thermal expansion coefficient is given as:

$$\alpha_v = \frac{1}{V} \cdot \frac{\partial V}{\partial T} \tag{25}$$

where V is volume entering the process at the condition of constant pressure and T is time. In solids, the pressure does not appreciably affect the size of an object. To an approximation with regard to measurement, the change in length of an object (which is linear dimension as opposed to volumetric dimension) due to thermal expansion is related to a temperature change by a linear expansion coefficient, which is defined as linear expansion due to a change in temperature with respect to initial length of an object, as shown in (26) and is expressed in K^{-1} [22]:

$$\alpha = \frac{\Delta l_t}{l_0 \cdot |T_2 - T_1|} \tag{26}$$

where α is thermal expansion coefficient, Δl_t – linear expansion; l_0 – initial length; T_1 is initial temperature and T_2 is final temperature.

Linear expansion denoted as Δl_t depends on initial length l_0 of an object and change in temperature. The thermal expansion coefficient can be obtained from data for either heating or cooling process. It does not depend on the shape of the object, temperature range or speed (i.e. how intense is the heating process). Linear expansion Δl_t of a solid measured by an interferometer is shown in (27) and is expressed in meters [4]:

$$\Delta l_t = 0.5 \cdot \lambda \cdot n \tag{27}$$

where n is number of interference fringes passed through an observation point of an interferogram, obtained with respect to optical path difference, and λ is wavelength of the laser as a source of coherent optical radiation. The measurement of temperature and the measurement of linear expansion must be done together. To determine a thermal expansion coefficient, the initial length of an investigated object must be well known.

Substituting (27) to (26) and knowing the initial and final temperature, it can be obtained:

$$\alpha = \frac{0.5 \cdot \lambda \cdot n}{l_0 \cdot |T_2 - T_1|} \tag{28}$$

The thermal expansion coefficient is constant for a given material. Thermal expansion coefficient can be determined for small linear expansion (i.e. achieved for a small change in temperature). Since temperature affects distribution of interference fringes, the experiment can be carried out in room temperature that facilitates temperature controllability. The block scheme of an experimental setup using a free space Michelson interferometer for measurement of thermal expansion coefficient of a copper rod is presented in Figure 6, and the practical arrangement is shown in Figure 7.

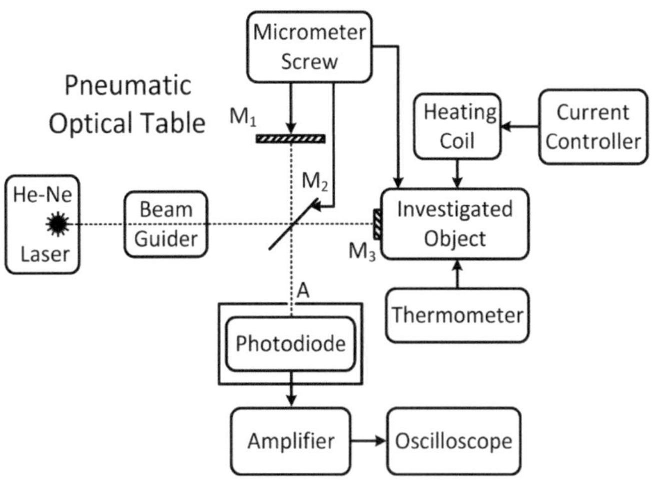

Figure 6. Schematic diagram of a free space Michelson interferometer for the measurement of thermal expansion coefficient for copper. M_2 – 50% mirror (beam splitter), M_1, M_3 – dielectric mirrors.

Figure 7. Schematic diagram of a free space Michelson interferometer for the measurement of thermal expansion coefficient for copper. 1, 2 – dielectric mirrors, 2 – beam splitter, 4 – observation plane (left), and detailed view on the investigated object (1) placed in a holder (2) with a micrometer screw (3), plastic isolation (4), heating coil, and a mirror in a mount (6) (right).

Optical path difference corresponding to one fringe is equal to value of half-wavelength. To obtain the value of linear expansion of a material, the value of half-wavelength is multiplied by the number of interference fringes. Data are gathered for monotonic increase in temperature. In general, temperature changes do not have to be monotonic because the key feature is to know an initial and final temperature, however, to facilitate fringe counting it is better to measure them at the condition of monotonic change in temperature. If the temperature change is not monotonic and both heating and cooling occur between an initial and final temperature, some fringes are measured several times and as a result the measured expansion is larger than it is in reality. Heating can be done by a coil, cooling is natural and is not accelerated by any cooling device. The measurement should be repeated many times and the measured values should be averaged in order to ensure the credibility of measurement.

The investigated object can be a metal rod with well-known initial length. A dielectric mirror should be placed on the tip of the rod and terminates the interferometer's arm. It is a high reflectance dielectric mirror created by the deposition of a thin layer of a dielectric material on a glass substrate. Eventual unexpected variation in room temperature could influence the path length and the distribution of fringes. For this reason, e.g. a gas burner cannot be used for heating.

We determine the thermal expansion coefficient for copper from the measured number of fringes, the initial rod's length and the temperature change. The number of light intensity maxima (i.e. the number of light fringes) multiplied by the value of half-wavelength is equal to the resultant expansion of the investigated copper rod. The result is averaged. For a well-known value of thermal expansion coefficient, the experimental setup can also work as a very accurate sensor of temperature changes that could for example be used in monitoring temperature stability in harsh environment. A similar application has been published [20]. Sample measured interferograms are shown in Figure 8.

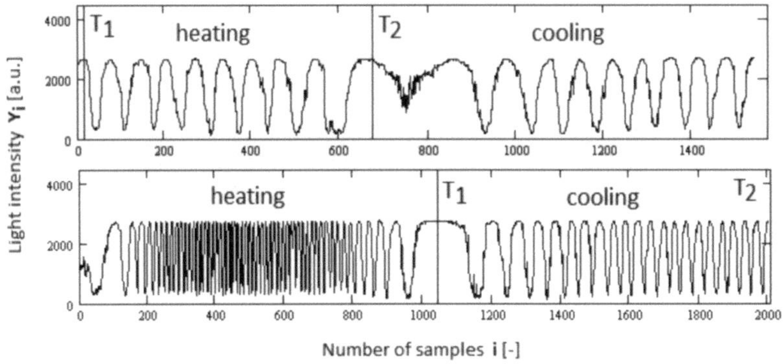

Figure 8. Sample data on measured light intensity over number of samples. T_1 is initial temperature and T_2 is final temperature of the measurement.

Table 1 shows measured data and the calculated thermal expansion coefficient. In Table 1, l_0 is initial length of the copper rod, T_1 is initial temperature, T_2 is final temperature, n is number of interference fringes, λ is He-Ne wavelength, Δl_t is linear expansion, and α is thermal expansion coefficient calculated using (24).

l_0	T_1	T_2	$\|T_2-T_1\|$	n	λ	Δl_t	α
(m) x10^{-3}	(K)	(K)	(K)	(-)	(m) x 10^{-9}	(m) x 10^{-6}	(K^{-1}) x 10^{-6}
82	296.7	317.5	20.8	105	632.8	33.222	19.47819
82	317.7	307.2	10.5	45	632.8	14.238	16.53659
82	312.8	301.3	11.5	49	632.8	15.504	16.44072
82	301.1	296.8	4.3	18	632.8	5.695	16.15201
82	296.7	302.7	6	27	632.8	8.543	17.36341
82	303.9	296.9	7	31	632.8	9.808	17.08780
82	296.8	301.0	4.2	20	632.8	6.328	18.37398
82	305.3	297.2	8.1	38	632.8	12.023	18.10178
82	323.3	303.3	20	92	632.8	29.109	17.74927
82	303.3	309.8	6.5	29	632.8	9.176	17.21501
82	305.2	303.0	2.2	10	632.8	3.164	17.53880
82	296.9	300.9	4	18	632.8	5.695	17.36341
82	297.3	315.6	18.3	80	632.8	25.312	16.86792
Mean							**17.40530**

Table 1. Calculation of a thermal expansion coefficient for copper.

The averaged value of thermal expansion coefficient, 17.4 $K^{-1}x10^{-6}$, corresponds well to the value 16.9 $K^{-1}x10^{-6}$ that can be found in [23].

http://dx.doi.org/10.5772/57149

4.2. Measurement of membrane vibration frequency using fiber optic Michelson interferometer

A schematic diagram of the experimental setup using a fiber optic Michelson interferometer for measurement of membrane vibration frequency is presented in Figure 9.

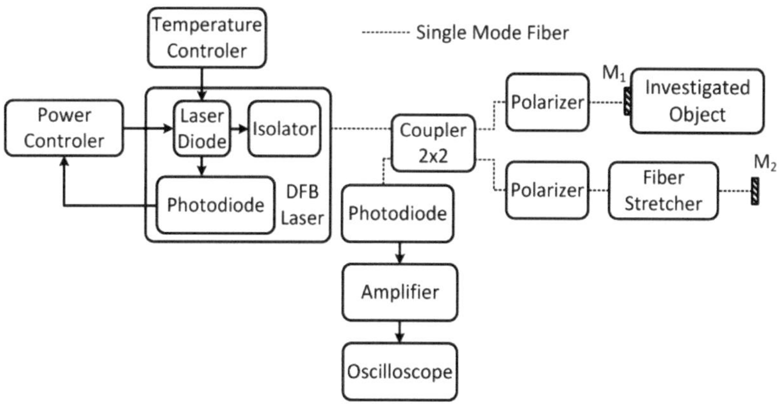

Figure 9. Schematic diagram of a fiber optic Michelson interferometer for the measurement of membrane vibration frequency. M_1, M_2 – pigtailed dielectric fiber mirrors.

The reference DFB laser beam is guided to the observation plane, going twice through the coupler. The measuring beam is guided to the investigated object, which is a loudspeaker mounted on a fiber mirror. As a result of the superposition of the two beams, interference fringes are observed. The interference fringes pass to a photo detector with an amplifier, and the maxima of signal intensity are displayed on an oscilloscope. Finally, data are exporter to PC.

The experimental setup should suppress ambient vibrations; otherwise the interference pattern is not stable. The experimental setup is placed on a hydro-pneumatic laboratory table and mounted by using special holders. Practical arrangement of the experimental setup is shown in Figure 10. The DFB laser operate at 1300 nm. Its temperature is monitored and controlled in order to provide a stable wavelength in time and narrow emitted bandwidth. Pigtailed mirrors, fiber couplers, polarizers are commercially available. The oscilloscope provides Fourier transform of the measured signal in order to measure the known frequency of vibrations of a loudspeaker.

The model of a high-speed and low amplitude physical quantity can be represented by the signal originated by the loudspeaker's membrane. The measurements were performed on a small 0.5W 8Ω, loudspeaker with the operation bandwidth 4 Hz – 1.8 kHz. The excitation signal is a sinus wave with constant amplitude. The excitation signal must not contain a constant voltage component, which causes the short circuit and can damage the loudspeaker. (To avoid this, a loudspeaker should be supplied through a capacitor). If a waveform generator is equipped with output impedance greater than tens of ohms, the loudspeaker should be connected to the generator with a resistor in series. However, too large resistance causes absorption of vibrations.

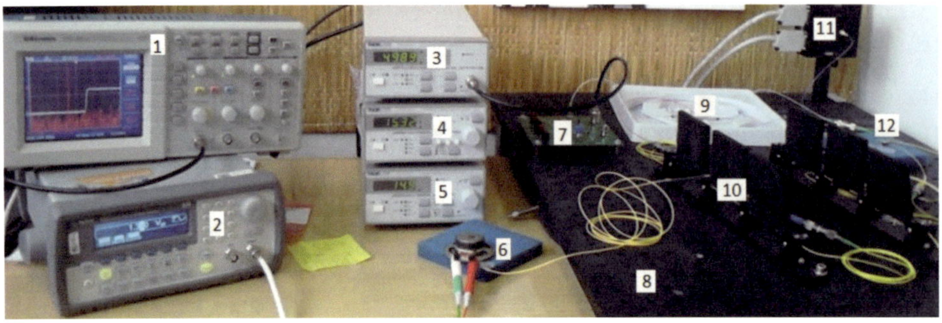

Figure 10. Practical arrangement of a fiber optic Michelson interferometer for the measurement of membrane vibration frequency. 1 – oscilloscope, 2 – waveform generator, 3 – electric amplifier, 4 – temperature controller, 5 – laser power controller, 6 – investigated object – a loudspeaker, 7 – photodiode, 8 – optical table, 9 – fiber coupler, 10 – polarizer, 11 – DFB laser in a special holder, 12 – pigtailed fiber mirror.

A fiber dielectric mirror mounted on the loudspeaker oscillates with its membrane. The information on its displacement is contained in the interference fringes. For rough measurement of vibration's frequency (but not precise determination of amplitude of vibrations), it is sufficient to place a loudspeaker close to the fiber mirror.

The change in light intensity on the detector's surface is caused by the vibration of the investigated membrane. The amount of fringes passing through the slit in front of an interference pattern corresponds to certain displacement of the membrane. Sample measured data is displayed in Figure 11.

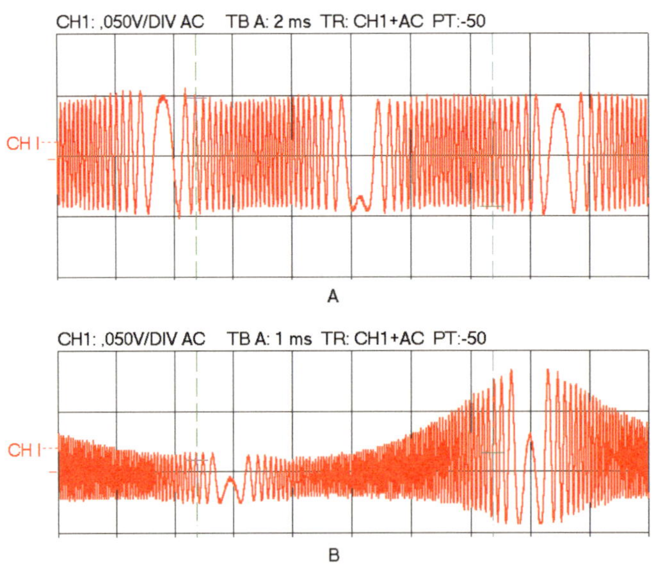

Figure 11. Sample data describing vibrations of a membrane. Frequency of oscillations: 75 Hz, number of maxima per period: 59 (A), and 99 Hz, number of light intensity maxima per period being 164 (B).

The broadened pitch between two light intensity maxima indicates the change to the direction of movement of the membrane. The amplitude of membrane's vibrations can be calculated by using (21). The circuit can measure the frequency of loudspeaker's sound (i.e. the frequency of membrane vibrations) by setting the Fourier transform on an oscilloscope. If the frequency is known, it can be a tool to verify that the setup works correctly. Then the frequency at a waveform generator connected to a loudspeaker should correspond well to the frequency measured and displayed on an oscilloscope. In addition, maximum vibration is measured for the resonant frequency.

5. Conclusions

The main advantage of the laser interferometry method is its precision, noncontact procedure and the application to objects specific for the small variation of signals measured on them. The experimental results confirmed the high accuracy of this optical method in this specific application area.

The measurements require performing many supplementary measurements. Every measurement is the source of inaccuracy. One has to consider potential inaccuracy sourcing from thermometers, length meters, lasers, as well as find the right solution for the detection and processing of fringes. Michelson interferometer requires mechanical stability especially in a free space arrangement to reduce spurious signals, strongly affecting the final result.

Considering the coherence length, the interferometer can be distributed to allow for measurement on far objects. In addition, the interferometer can be miniaturized or built as a fiber optic based sensor.

Acknowledgements

This work was supported by the Ministry of Interior of the Czech Republic grant under project VG20102015053, "Guardsense" and by the by the CTU grant under project SGS13/201/OHK3/3T/13.

Author details

Michal Lucki*, Leos Bohac and Richard Zeleny

*Address all correspondence to: lucki@fel.cvut.cz

Department of Telecommunications Engineering, Faculty of Electrical Engineering, Czech Technical University in Prague, Prague, Czech Republic

References

[1] Malak M., Marty. F., Nouira H., Salgado J. A., Bourouina T. All-silicon interferometric optical probe for noncontact Dimensional measurements in confined Environments. Proceedings of the IEEE 25th International Conference on Micro Electro Mechanical Systems; 2012 628-631. http://ieeexplore.ieee.org/xpl/articleDetails.jsp?tp=&arnumber=6170266&queryText%3DAll-silicon+interferometric+optical+probe+for+noncontact+Dimensional+measurements+in+confined+Environments (accessed 4 September 2013).

[2] Luhs W. Experiment 10 Michelson interferometer. Eschbach: Laserzentrum FH Münster University of Appl. Sciences; 1995 revised 2003. http://repairfaq.ece.drexel.edu/sam/MEOS/EXP10.pdf (accessed 4 September 2013).

[3] Mokryy O., Koshovyy V., Romanyshyn I., Sharamaga R. Stabilized detection scheme of surface acoustic waves by Michelson interferometer. Optica Applicata 2010; 40(2) 449-458. http://www.if.pwr.wroc.pl/~optappl/pdf/2010/no2/optappl_4002p449.pdf (accessed 4 September 2013).

[4] Scholl R., Liby B. W. Using a Michelson Interferometer to Measure Coefficient of Thermal Expansion of Copper. The Physics Teacher 2009; 47(5) 316-318. http://physlab.lums.edu.pk/images/2/2c/Scholl_liby.pdf (accessed 4 September 2013).

[5] Tol J., A Mach-Zehnder-interferometer-based low-loss combiner. Photonics Technology Letters 2001; 13(11) 1197-1199. http://ieeexplore.ieee.org/xpl/articleDetails.jsp?tp=&arnumber=959362&queryText%3DTol+J.%2C+Huig+G.%2C+Yong+L.+A+Mach-Zehnder-interferometer-based+low-loss+combine.+Photonics+Technology+Letters (accessed 4 September 2013).

[6] Fang X. A variable-loop Sagnac interferometer for distributed impact sensing. Lightwave Technology 1996; 14(10) 2250-2254. http://ieeexplore.ieee.org/xpl/articleDetails.jsp?tp=&arnumber=541215&queryText%3DA+variable-loop+Sagnac+interferometer+for+distributed+impact+sensing (accessed 4 September 2013).

[7] Hariharan P. Basics of interferometry. Sydney: School of Physics, University of Sydney, Elsevier; 2007

[8] Zagar B. A laser-interferometer measuring displacement with nanometer resolution. IEEE Transactions on Instrumentation and Measurement 1994; 43(2) 332-336. http://ieeexplore.ieee.org/xpl/articleDetails.jsp?tp=&arnumber=293440&queryText%3DA+laser-interferometer+measuring+displacement+with+nanometer+resolution (accessed 4 September 2013).

[9] Iyer S., Coen S., Vanholsbeeck F. All-fiber optical coherence tomography system incorporating a dual fiber stretcher dispersion compensator. Proc. SPIE 7004, 19th International Conference on Optical Fibre Sensors; 2008, 700434-1–700434-4. http://

proceedings.spiedigitallibrary.org/proceeding.aspx?articleid=786962 (accessed 5 September 2013).

[10] Winter A., Schlarb H., Schmidt B., Ilday F.O., Jung-Won Kim Chen J., Grawert F.J., Kartner F.X. Stabilized Optical Fiber Links for the XFEL. Proceedings of the Particle Accelerator Conference; 2005 2589-2591. http://ieeexplore.ieee.org/xpl/articleDetails.jsp?arnumber=1591192 (accessed 4 September 2013).

[11] Kersey A. D., Marrone M. J., Dandridge A., Tveten, A. B. Optimization and Stabilization of Visibility in Interferometric Fiber-optic Sensors Using Input-Polarization Control. Journal of Lightwave Technology 1988; 6 (10) 1599-1609.

[12] Diniz P., Da Silva E., Netto S. Digital Signal Processing: System Analysis and Design. Cambridge: Cambridge University Press; 2002. http://assets.cambridge.org/97805217/81756/copyright/9780521781756_copyright.pdf (accessed 4 September 2013).

[13] Narayanamurthy C. S. Analysis of the localization of Michelson interferometer fringes using Fourier optics and temporal coherence Eur. J. Phys. 2009; 30 147-155. http://iopscience.iop.org/0143-0807/30/1/015/pdf/ejp9_1_015.pdf (accessed 4 September 2013).

[14] Alzahrani K., Burton D., Lilley F., Gdeisat M., Bezombes F. Automatic Absolute Distance Measurement with One Micrometer Uncertainty Using a Michelson Interferometer. Proceedings of the World Congress on Engineering; 2011, Vol 2. http://core.kmi.open.ac.uk/display/1019144 (accessed 4 September 2013).

[15] Boyce W., Di Prima R. Elementary Differential Equations. New York: John Wiley & Sons, Inc.; 1986.

[16] Cheng K., Shang X., Cheng X. The Digitization of Michelson Interferometer. Proc. of 2011 International Conference on Electronics and Optoelectronics; 2011 V3142-V3145.

[17] Byeong H. L., Young H. K., Kwan S. P., Joo B. E., Myoung J. K., Byung S. R., Hae Y. Ch. Interferometric Fiber Optic Sensors. Sensors 2012; 12 2467-2486. http://www.mdpi.com/1424-8220/12/3/2467 (accessed 4 September 2013).

[18] Yung-Cheng W., Lih-Horng S., Chung-Ping Ch. The Comparison of Environmental Effects on Michelson and Fabry-Perot Interferometers Utilized for the Displacement Measurement. Sensors 2010; 10 2577-2586. http://www.mdpi.com/1424-8220/10/4/2577 (accessed 4 September 2013).

[19] Acernesea F., De Rosab R., Garufi F., Romano R., Barone F. A Michelson interferometer for seismic wave measurement: theoretical analysis and system performances. Proc. SPIE 6366 Remote Sensing for Environmental Monitoring, GIS Applications, and Geology VI; 2006. http://spie.org/x648.html?product_id=687907 (accessed 4 September 2013).

[20] Li X., Lin S., Liang J., Zhang Y., Ueda O. Fiber-Optic Temperature Sensor Based on Difference of Thermal Expansion Coefficient Between Fused Silica and Metallic Ma-

terials. IEEE Photonics Journal 2012; 4(1) 155-162. http://ieeexplore.ieee.org/xpl/articleDetails.jsp?tp=&arnumber=6112705&queryText%3DFiber-Optic+Temperature+Sensor+Based+on+Difference+of+Thermal+Expansion+Coefficient+Between+Fused+Silica+and+Metallic+Materials

[21] Kezmah M., Donlagic D., Lenardic B. Low Cost Security Perimeter Based on a Michelson Interferometer. Proc of IEEE Sensors; 2008 1139-1142. http://ieeexplore.ieee.org/xpl/articleDetails.jsp?tp=&arnumber=4716642&queryText%3DLow+cost+security+perimeter+based+on+a+Michelson+interferometer (accessed 4 September 2013) (accessed 4 September 2013).

[22] Cverna F., editor. ASM International Materials Properties Database Committee. Thermal Properties of Metals. ASM Materials Data Series; 2002. http://www.asminternational.org/content/ASM/StoreFiles/ACFAAB7.pdf (accessed 4 September 2013) (accessed 4 September 2013).

[23] Chapter12. In: Lide D. CRC Handbook of Chemistry and Physics, Boca Raton: CRC Press; 2003. http://www.hbcpnetbase.com/ (accessed 4 September 2013).

Chapter 2

Optical Fibre Sensors Based on UV Inscribed Excessively Tilted Fibre Grating

Chengbo Mou, Zhijun Yan, Kaiming Zhou and
Lin Zhang

Additional information is available at the end of the chapter

http://dx.doi.org/10.5772/57146

1. Introduction

In-fibre inscription of grating structures in the core of an optical fibre was firstly reported in 1978 by Hill [1]. Being a promising device as narrow band reflector, the fibre Bragg gratings (FBGs) have drawn a lot of attentions in the field of optical communication at that time. However, the functionality of FBGs as sensors has been only recognised after a decade of the invention of the device which can inscribe FBG with resonant wavelength independent of the writing laser wavelength [2]. Since then, the research of FBG based sensors has grown tremendously [3]. The techniques using diffractive optical element to fabricate FBG have put the field into a more commercial way as reproductive of identical FBGs is possible [4]. FBGs are then found a range of applications in sensing field such as strain, temperature, curvature, loading, displacement etc. In 1996, a new type of fibre grating device which is called long period fibre grating (LPG) was demonstrated which has superior temperature sensitivity while possessing refractive index (RI) responsivity [5]. Both FBGs and LPGs have shown significant role in the optical sensing domain. They have been utilised directly or functionalised or integrated with other structures to show functionality in various sensing applications.

Another class of in-fibre gratings is the grating structure with tilted grating planes which called tilted fibre gratings (TFGs). Such a type of gratings is capable of couple the core propagating mode into strong cladding modes. In terms of the tilted angles, such gratings can be divided into three types namely small angle (<45°) TFG, 45° TFG and excessively (>45°) TFG (ETFG). The small angle TFGs were originally used as mode coupler which taps the light out from the fibre core area [6]. Recently, such gratings have shown strong potential in sensing field [7-10]. When incorporated with metal coating, such gratings also exhibit great potential for refractive index sensing based on surface plasmon resonance [11]. 45°-TFG was initially demonstrated as a polarisation dependent loss equaliser [12] and later as an in-fibre polariser [13]. The ETFG is a new class of fibre gratings which was first demonstrated in 2006 by Zhou *et al* [14]. Since then, the ETFGs have shown great capability as a novel kind of fibre sensors. This chapter will review the recent development of ETFGs as various optical sensors. The chapter will be organised in three main parts: first part (sections 2) gives a general introduction and fundamental background on fibre gratings with a particular emphasis on the ETFG; second part (section 3) describes the inscription and characterisation of ETFG; third part (sections 4-9) discusses ETFG based sensors and fibre laser sensing systems including strain, twist [15, 16], loading[17], refractive index (RI) and liquid level sensing [18].

2. Background of fibre gratings

Light coupling in a non-tilted fibre grating can be well illustrated by ray tracing as shown in Figure 1. For an FBG, the mode coupling occurs at resonant wavelength where the forward propagating mode reflects into an identical backward propagating mode (Figure 1a). While for an LPG, the mode coupling occurs close to wavelength at which a forward propagating core mode is strongly coupled into co-propagating cladding modes (Figure 1b). For TFGs, the mechanism of light coupling can also be described by ray tracing method as shown in Figure 2.

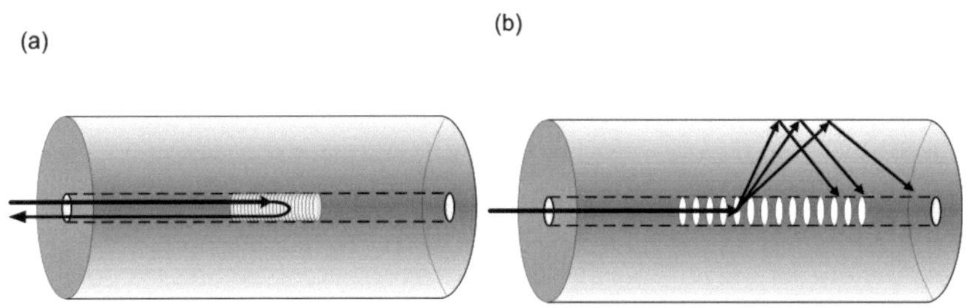

Figure 1. Schematic of light coupling of (a) a FBG showing light coupled from forward propagating core mode to backward propagating cladding mode; (b) a LPG showing light coupled from forward propagating core mode to forward propagating cladding modes.

As can be seen from Figure 2, when the grating tilted angle is smaller than 45°, the grating is capable of coupling forward propagating core mode into backward propagating cladding modes (Figure 2a). At 45°, as a unique case, the core mode will be coupled into radiation mode normal to the fibre axis (Figure 2b). When the tilted angle is larger than 45°, like LPGs, the ETFGs are capable of coupling the forward propagating core mode into forward propagating cladding modes, but to the high order ones (Figure 2c). The strongest light coupling occurs at the resonant wavelength where the phase matching condition $\lambda_{co\text{-}cl}=(n_{c0}\pm n_{cl,m})\ \Lambda/\cos\theta$ is satisfied, where n_{c0} and $n_{cl,m}$ are the effective mode refractive indices of the core mode and the mth cladding mode, Λ is the grating period and θ is the tilted angle of the grating structure. The mode coupling mechanism can be well understood by the phase matching condition. We hereby define the following wave vector relationship for mode coupling in a fibre grating

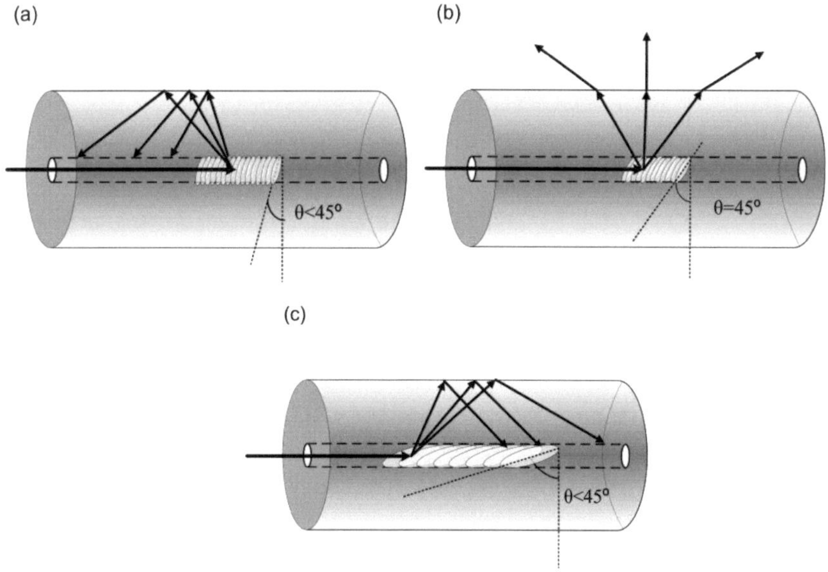

Figure 2. Schematic of light coupling of TFGs with tilted angle (a)< 45°; (b)=45°;(c)>45°.

which is commonly regarded as the phase matching condition$K_{out} = K_{in} + K_G$. All K described in this section are vectors. $K_{in} = \frac{2\pi}{\lambda} \cdot n_{co}$is the wave vector of the incident light and $K_G = \frac{2 \cdot \pi}{\Lambda_G}$ is the grating vector. The phase matching condition of a fibre grating can then be described in a vectorial plane in Figure 3 and Figure 4. For the case of FBG mode coupling, as shown in Figure 3a, the relationship $K_{out} = K_{in} = \frac{2 \cdot \pi}{\lambda} \cdot n_{co}$ applies as an FBG structure will couple the light from a forward propagating core mode into an identical backward propagating core mode. For the case of LPG mode coupling, as shown in Figure 3b, the grating can couple the incident light into forward propagating cladding modes with $K_{out} = \frac{2\pi}{\lambda} \cdot n_{cl}$ indicating the cladding modes.

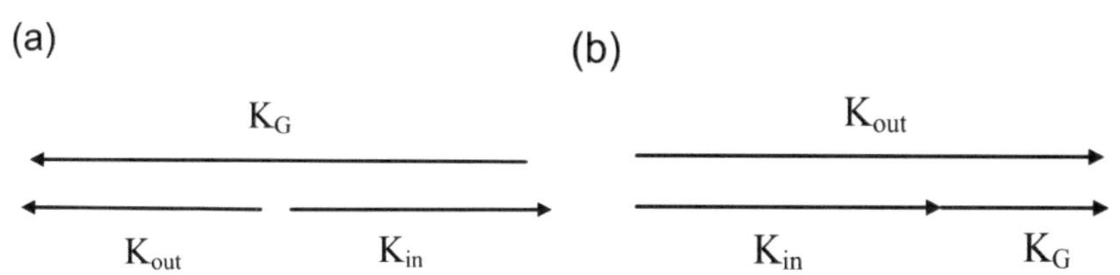

Figure 3. Vectorial descriptions of phase matching conditions of (a) FBG and (b) LPG.

For TFGs, as the grating has the ability to couple the forward propagating core mode into radiation mode, it is hence to have $K_{out} = \frac{2 \cdot \pi}{\lambda} \cdot n_{clad}$which is similar to LPG. With the condition

$n_{co} \cong n_{clad}$, the following relationship $K_{in} \cong K_{out}$ therefore applies. Hence, the phase matching condition of TFGs can be depicted in the vector plane which is shown in Figure 4, where θ indicates the tilted angle of the grating with respect to the fibre axis. In Figure 4a, we can simply infer that when the tilted angle is minimised to zero, the phase matching illustration evolves into the standard FBG condition from which a forward propagating mode has been coupled into an identical backward propagating mode via Bragg diffraction. Figure 4b shows the special case of 45°-TFG which is capable of coupling out light perpendicular to the fibre axis or incident beam propagation direction. While Figure 4c shows the mechanism of an incident beam couples into a forward propagating mode through an excessively titled grating structure. Although the phase matching condition gives very good approximation for interpretation of mode coupling mechanism inside the TFGs, it does not involve the polarisation effect which is actually one of the key properties of the TFGs.

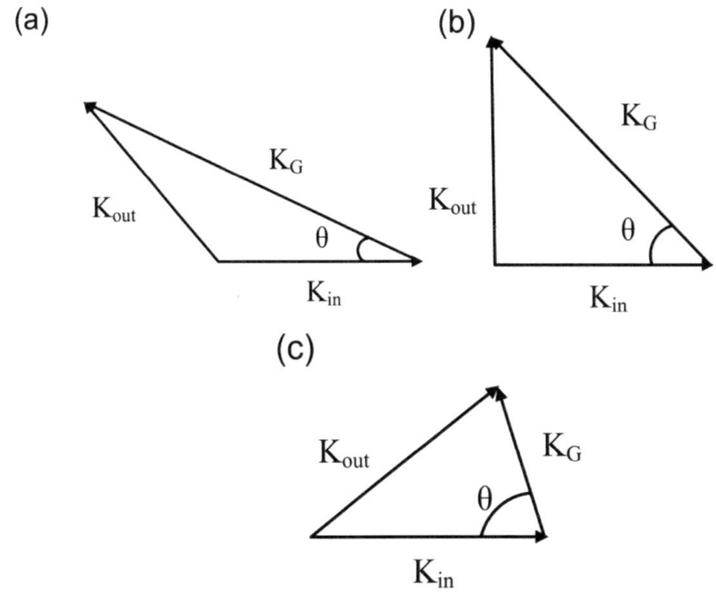

Figure 4. Vectorial description of phase matching conditions for TFGs with titled angles at (a) < 45°, (b) = 45°and (c) > 45°.

Due to their large tilted angle induced strong asymmetry to the fibre geometry, ETFGs exhibit polarisation dependent mode splitting which features with pairs of peaks corresponding to two orthogonal polarisation modes. We can therefore identify an equivalent fast-axis and slow-axis similar to the conventional polarisation maintaining (PM) fibre structure as shown in Figure 5. It is this distinctive polarisation mode splitting mechanism makes ETFGs as ideal loading [17] and twisting sensors [16] based on their polarisation property and as refractive index sensors utilising intrinsic sensitivity of the high order modes to surrounding medium.

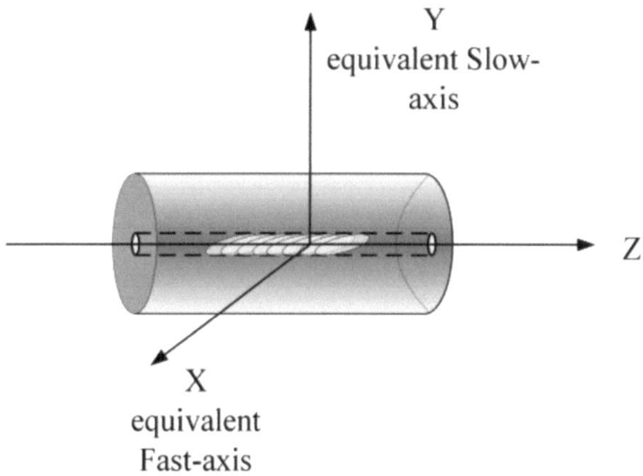

Figure 5. Schematic illustration of an ETFG structure with two assigned orthogonal polarisation axes.

3. Fabrication and spectral properties of ETFG

As illustrated in Figure 6, a TFG can be inscribed either by tilting the mask with respect to the fibre axis (Figure 6a), or by using a mask with tilted pitches (Figure 6b). As an alternative approach, one can inscribe such gratings by tilting the fibre about its axis orthogonal to the plane defined by the two interfering UV beams in a two-beam holographic fabrication system (Figure 6c). A commercial argon ion UV laser is employed to inscribe ETFG in hydrogenated standard telecom fibre (SMF28). Similar to standard FBG fabrication, we have adopted mask scanning technique for ETFG inscription due to high reproducibility and fine control of the grating devices. A commercial amplitude mask with 6.6 μm period was purchased for ETFG inscription ensuring the spectral response residing within a broad range from 1200 to 1700 nm. The schematic UV inscription setup is shown in Figure 7a. A typical microscopic image of an ETFG is shown in Figure 7b demonstrating the slanted grating fringes at ~78°.

A broadband light source (BBS), a polariser and a polarisation controller (PC) are utilised to examine the spectral properties of ETFG through an optical spectrum analyser (OSA). A typical measurement schematic setup is illustrated in Figure 8.

Figure 9a shows the optical spectrum of a typical ETFG from 1200 to 1700 nm, exhibiting unique paired loss peaks due to polarisation mode splitting, when probed using unpolarised BBS. When polarised light with proper polarisation state is launched as a probe, only one set of split modes will be excited and the other set disappears. As can be seen from Figure 9b, either the equivalent fast- (blue dash-dotted line) or the slow-axis (red-dashed line) mode can be fully excited or eliminated with polarised light.

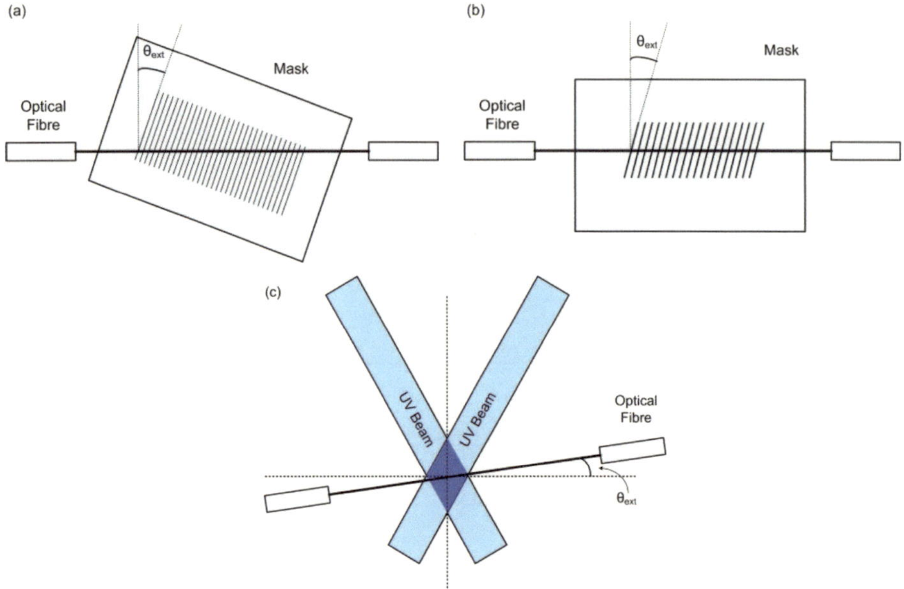

Figure 6. (a) and (b) mask scanning (c) two-beam holographic technique for TFG fabrication.

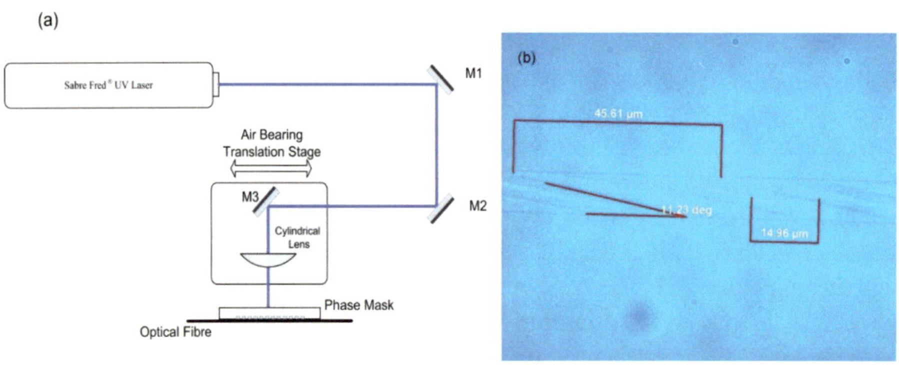

Figure 7. (a) Schematic of UV mask scanning technique for ETFG fabrication; (b) a typical microscopic image of an ETFG (Reprinted from Optics Communication Vol.305 pp271-275, Copyright (2013) with permission from Elsevier).

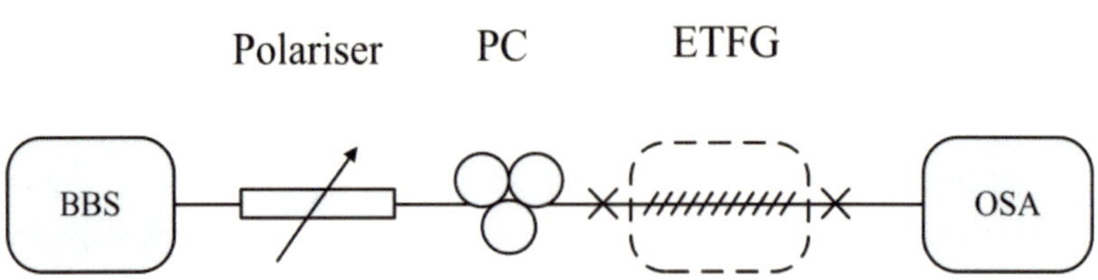

Figure 8. Typical measurement schematic for characterising ETFG.

http://dx.doi.org/10.5772/57146

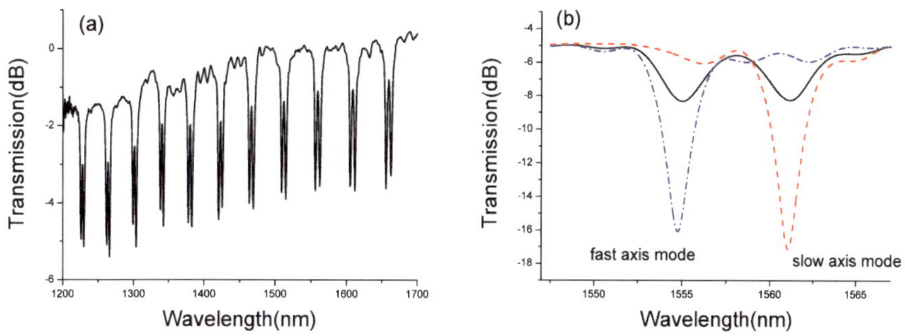

Figure 9. (a) Optical spectrum of a typical ETFG from 1200 nm to 1700 nm; (b) the spectra of one zoomed loss peak pair when polarised light probing the ETFG (Reprinted from Optics Communication Vol.305 pp271-275, Copyright (2013) with permission from Elsevier).

4. Thermal responsivity of ETFG

The ETFGs have shown a thermal responsivity as low as 3.3 pm/°C in 1200 nm range [19]. For the application of optical sensors, the thermal responsivity in the 1550 nm range is of interest. The thermal responsivity of the ETFG has been examined by mounting the grating on a peltier device based heat exchange board using a commercial temperature controller while monitoring the transmission spectrum change with elevated temperature. We studied two pairs of loss peaks of the ETFG around 1560 nm and 1610 nm individually. Figure 10 plots the wavelength shift of the two paired loss peaks when the temperature of the grating increases from 20 °C to 80 °C with a step of 10 °C. Because a polariser and a PC has been used in the experiment, extra insertion loss is therefore induced to the system. Moreover, the BBS has a low power response at the interested wavelength range. Hence, the measured intensity of the loss peaks almost reach the sensitivity limit of the OSA. While the resolution of the OSA used in the experiment was limited to 0.02 nm, the errors in the experiment in terms of wavelength change is 0.04 nm which is shown in the error bars. Figure 10 shows the thermal responsivities of the two paired loss peaks have a quasi linear relationship. The thermal responsivities of the fast- and slow-axis modes around 1560 nm are 4.5 pm/°C and 5.5 pm/°C (Figure 10a) while around 1610 nm are 4.5 pm/°C and 7.5 pm/°C (Figure 10b). It can be clearly seen that the thermal responsivity of ETFG depends on the mode orders, this is quite similar to the thermal behaviour of normal LPGs [20]. We have also found that the thermal responsivity of slow-axis mode is slightly higher than that of the fast-axis mode. Furthermore, compared to the conventional LPG [20], ETFG shows a much lower temperature sensitivity. Therefore, the ETFG could be an ideal optical sensor without compulsory temperature compensation scheme.

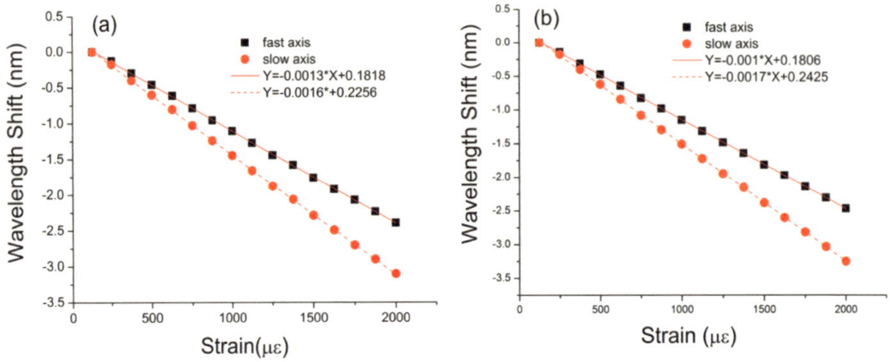

Figure 10. Wavelength shifts of two paired loss peaks of the ETFG against the temperature change in the ranges around (a) 1560 nm and (b) 1610 nm (Reprinted from Optics Communication Vol.305 pp271-275, Copyright (2013) with permission from Elsevier).

5. ETFG based strain sensing

To evaluate the strain responsivity of the ETFG, the grating fibre was mounted in a homemade fibre stretcher where the fibre was clamped on two metal block holders with a fixed distance, one of which being fitted with a precision translational micrometer driver. By moving the micrometer driver, the fibre was then stretched therefore inducing the strain from 0 to 2000 με. Figure 11 depicts the wavelength shift against applied strain for two paired loss peaks. The figure shows a linear relationship between the wavelength change and the applied strain. It can be seen that the strain responsivities of the fast- and slow-axis modes around 1530 nm are 1.3 pm/με and 1.6 pm/με (Figure 11a) while are 1 pm/με and 1.7 pm/με (Figure 11b) in the region around 1610 nm. We notice that the strain responsivity of the ETFG is slightly higher than FBG [3]. The strain responsivity of the fast-axis mode is generally higher than that of the slow-axis mode. It also worth to notice that the resonant wavelength of the ETFG has blue shift while it is under tensile strain. This is in contrast to the FBG strain response, however, corresponds very well to an LPG with relatively small period [20].

Although the passive detection of wavelength shift can offer smart sensing solutions, the systems are still subjected to complexity. Normally, in this case, an additional light source is necessary. Active strain sensors using fibre laser configuration provide an alternative measurement method with higher signal to noise ratio while having a less complicated system. Moreover, most of the passive sensing systems rely on optical spectrum domain signal demodulation from which the cost is high. Time domain signal demodulation offering low system cost has been reported through integrating a conventional LPG in a linear laser cavity [21]. In the following section, we describe the demonstration of a fibre laser strain sensor incorporating an ETFG. Low cost time domain signal demodulation can be achieved by monitoring the built-up time of the modulated laser cavity. The built-up time of the laser system is subject to the loss change of the cavity in a modulated laser system when gain and pump condition is constant. Because the ETFG has spectral loss bands, when it is subjected to mechanical strain in the laser cavity, the loss band will then shift accordingly, so that the cavity loss is related with applied strain and the built-up time of laser system will change correspondingly. Therefore the strain can be detected by monitoring the built-up time of the laser system.

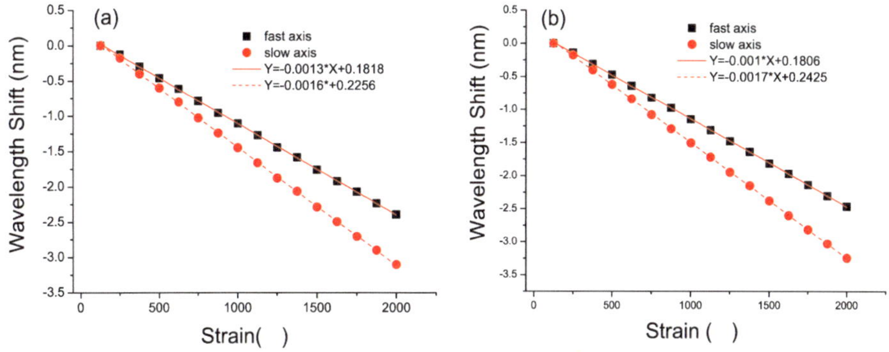

Figure 11. Wavelength shifts of two paired loss peaks of the ETFG against the temperature change in the ranges around (a) 1530 nm and (b) 1620 nm.

The setup for the proposed fibre laser strain sensor system is shown in Figure 12a. The ETFG is again mounted in a home made fibre stretcher. The pump LD is modulated by a square wave through a standard function generator at 5 Hz. The laser output is connected to a low noise photodiode, and the built-up time is measured via a standard two-channel digital oscilloscope. Figure 12b shows a typical oscillation trace of the laser oscillation built-up process.

In the experiment, by tuning the micrometer, we can obtain equivalent strain from 0 με to 2000 με applied on the ETFG. When the ETFG is at 0με, this corresponds to the maximum loss in the laser cavity which gives out the longest built-up time. While the strain increases, the loss bands will have a blue shift thus decreasing the cavity loss which results in a shorter built-up time of the system. A typical output spectra change is illustrated in Figure 13a.

The system was firstly set the lower and upper level modulation pump at constant values of 14.6 mW and 112.4 mW individually. The built-up time was then measured for the strain applied on the ETFG with an increment of 100 με. The absolute built-up time change against the applied strain on the ETFG is depicted in Figure 13b. From Figure 13b it can be seen that, initially when the ETFG is under no strain, the laser cavity suffers the maximum loss therefore exhibiting the largest built-up time. The experiment was then repeated for a different lower pump level at 24 mW. It can also be found that this laser strain sensor system is subject to the saturation of the applied strain as the system is in different lower pump levels. For a better understanding of the experimental results, it has been re-plotted from which linear range of the sensor response counts for in Figure 13b (i.e. lower pump level 24 mW) and Figure 13c (i.e. lower pump level 14.6 mW). It indicates, when the lower pump level is 14.6 mW, the sensor system can measure strain from 0 με to 1000 με with a linear responsivity of ~500 ns/με. This is far beyond the resolution of a standard oscilloscope. The sensor then reaches its saturation point when the applied strain is over 1000 με. For lower pump level at 24 mW, the sensor is only capable of measuring strain from 0 με to 500 με with a linear strain responsivity of ~349 ns/με.

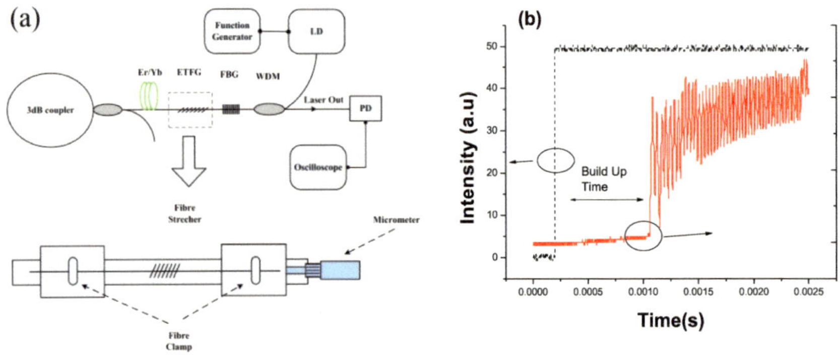

Figure 12. (a) Schematic diagram of the ETFG based fibre laser torsion sensor system; (b) Typical built-up time trace of the fibre laser observed on a digital oscilloscope. Modulation signal is shown in black dotted line; laser output signal is shown in red solid line.

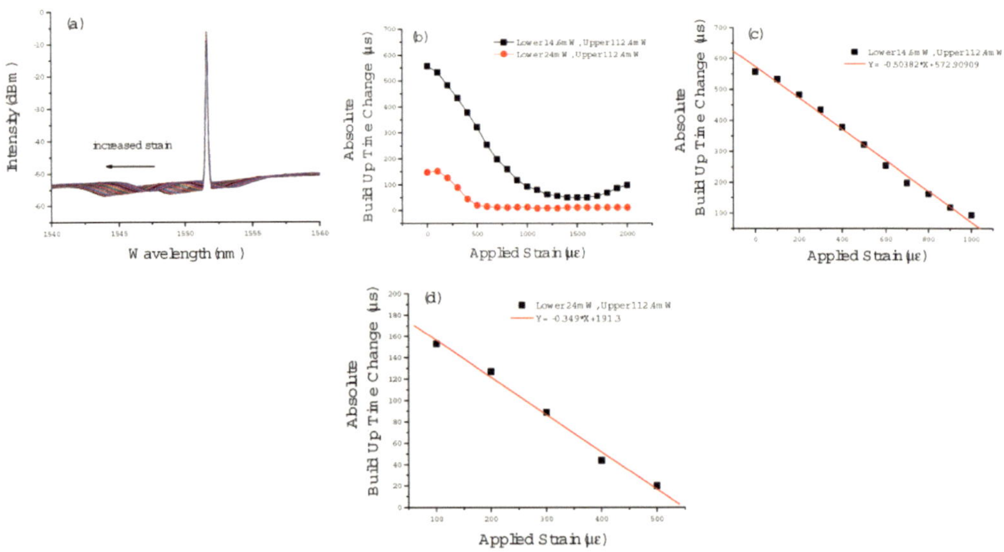

Figure 13. (a) Typical output spectrum change when the 79°-TFG is under strain; (b) Laser oscillation built-up time against applied strain on 79°-TFG for two different lower modulation pump power levels, and re-plotted separately for (c) lower pump level 14.6 mW from 0 to 1000 με (d) lower pump level 24 mW from 0 με to 500 με.

6. ETFG based twist sensing

As the ETFG is polarisation dependent, when probed with polarised light, the polarisation direction of the light will alternate accordingly if the fibre is under twist. The schematic of ETFG twist sensor is shown in Figure 14. In Figure 14, the broadband light is polarised through a fibre pigtailed linear polariser. The PC is then employed to alternate the state of polarisation of the input light to excite either fast-axis or slow-axis mode of the ETFG. One side of the ETFG is clamped on a stage using a metal block while the other side of the ETFG is fed into the OSA through a fibre rotator. The length between the fibre clamp and the fibre rotator is defined as L. A small tension was then applied to the fibre in order to eliminate the axial strain and bending effects, which may induce measurement uncertainty. Before the twist measurement commenced, the zero degree of rotation was normalised to a state that only fast-axis mode is fully excited by adjusting the PC. The twist was then applied to the grating in clockwise direction from 0° to 180° with 10° increment. The resultant transmission spectra evolution is depicted in Figure 15. From Figure 15 one can clearly see that when the ETFG is under twist, the strength of fast-axis mode increases while that of the slow-axis mode decreases. More importantly, the fast-axis mode diminished completely when the twist angle is 180°. A *vice versa* evolution was also observed when the twist was applied in the anti-clockwise direction from 0° to 180° between the fast-axis and slow-axis mode.

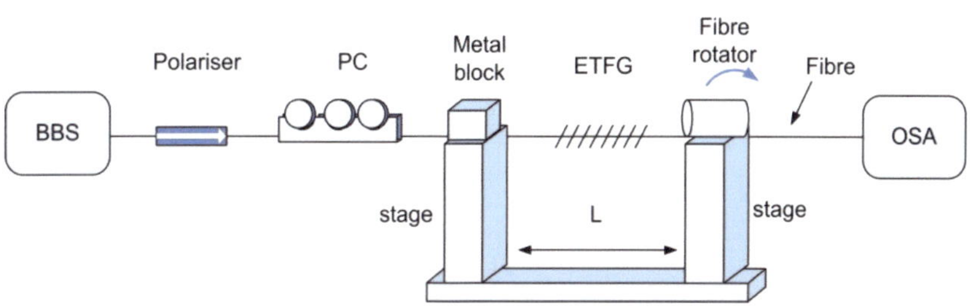

Figure 14. Schematic description of ETFG twist sensor system using a BBS.

In order to make the system more integrated, we carefully spliced the ETFG with a 45°-TFG make the system a compact all-fibre grating based system. While the polarising axis of the 45°-TFG matches either the fast-axis or slow-axis, the corresponding mode will be excited so that the necessity of PC adjustment is removed. To further lower down the cost of the system, single wavelength laser (SWL) was employed as a light source to which the laser line matches either the fast-axis or slow-axis mode. Therefore, while the twist was applied on the ETFG, the power variation can be recorded using a low cost power detector rather than an expensive OSA. The experimental setup of this improved low cost system is shown in Figure 16.

To perform the twist experiment, the SWL is set at the wavelength matching either the fast-axis or slow-axis mode as shown in Figure 17. The zero position of the sensor was calibrated by optimising the fibre rotator which has a minimum transmission power. The twist was then applied again from 0° to 180° with an elevation step of 10° for both fast-axis and slow-axis mode.

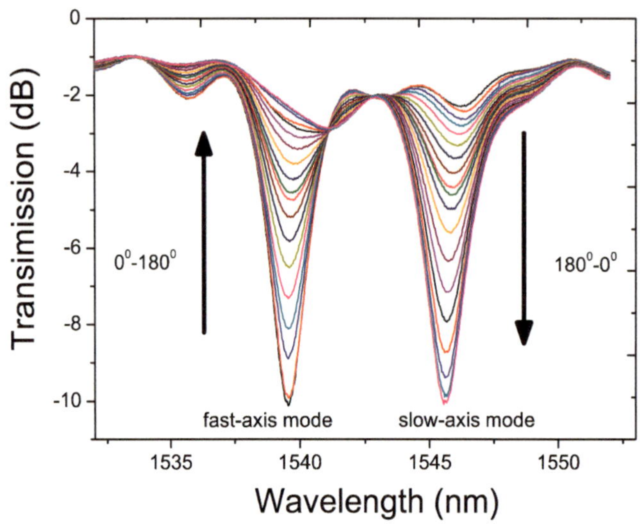

Figure 15. Spectral evolution of ETFG under twist in clockwise direction from 0° to 180° with an increment of 10°.

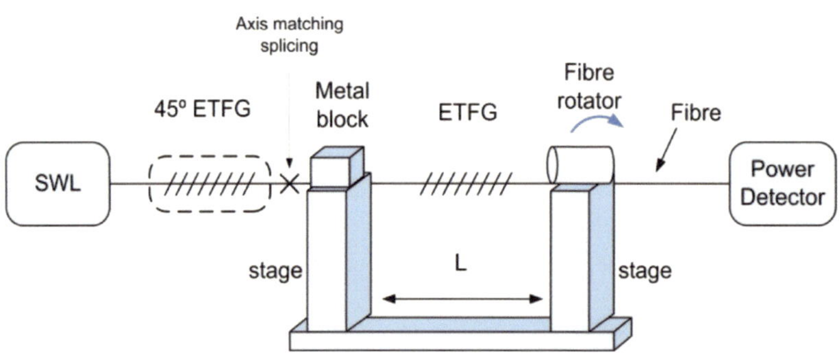

Figure 16. Schematic experimental setup of all-fibre grating based twist sensor system.

The result of the ETFG based twist sensor using power detection method is shown in Figure 18. In Figure 18a, it demonstrates that from 0° to 180° the transmission power of the fast-axis mode increases from -10 dBm to -2 dBm while that of the slow-axis mode is *vice versa*. One may identify a linear range of the sensor at the position of 90°±30°. Thus the sensitivity of the sensor is 0.1 dBm/(rad/m) and 0.24 dBm/(rad/m) for the slow-axis and fast-axis mode respectively. In Figure 18b, it shows similar results when detected with a photodetctor resulting in a voltage change from 0 to 3000 mV. The corresponding linear range gives out a sensitivity of 102.4 mW/(rad/m) and 101.8 mW/(rad/m) for the slow-axis and fast-axis mode individually. It can be seen clearly that when using a photodetector, the linearity is better than using a power meter. Also, the sensitivity is slightly better when using a photodetector. The twist sensitivity of both fast-

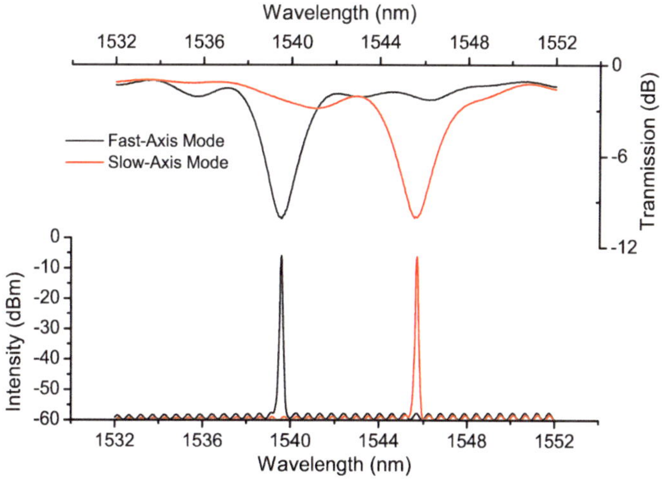

Figure 17. The upper plot is the transmission spectra of the ETFG; the lower one is the output spectra of a SWL set at the wavelength matching either fast-axis or slow-axis mode.

axis and slow-axis mode is quite similar. Therefore, in the real application, either mode can be used. The successful demonstration of using photodetector may provide a mechanism that the signal could be potentially transmitted through wireless control and remote monitoring.

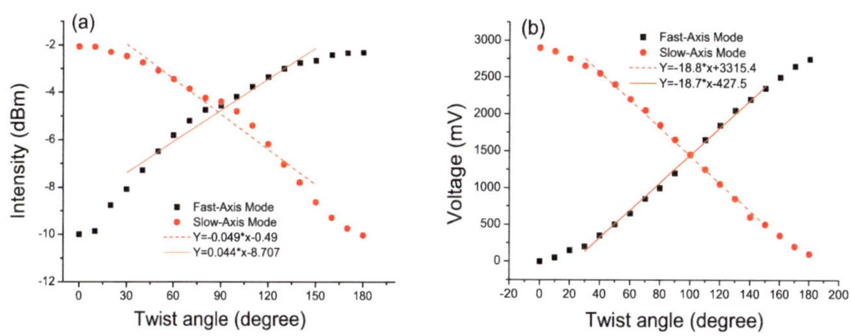

Figure 18. Transmission power variation against twist angle for fast-axis and slow-axis modes measured through low-cost power detection methods: (a) using a power meter and (b) a photodetector.

Similar to strain sensing, the ETFG can also be incorporated into a fibre laser to form a fibre laser based twist sensor system where time domain signal demodulation technique can be applied. The system setup is quite similar to the ETFG based fibre laser strain sensing system, the only difference is to replace the fibre clamps for strain by a set of fibre rotator. The setup for the ETFG based fibre laser twist sensor system is shown in Figure 19. The ETFG itself is a polarisation dependent loss filter, when an ETFG is inserted to the laser cavity and subject to twist, the intracavity loss will change accordingly, thus affecting the laser built-up time. Based on this principle, the twist experienced by the ETFG can therefore be monitored by measuring the built-up time of the laser cavity.

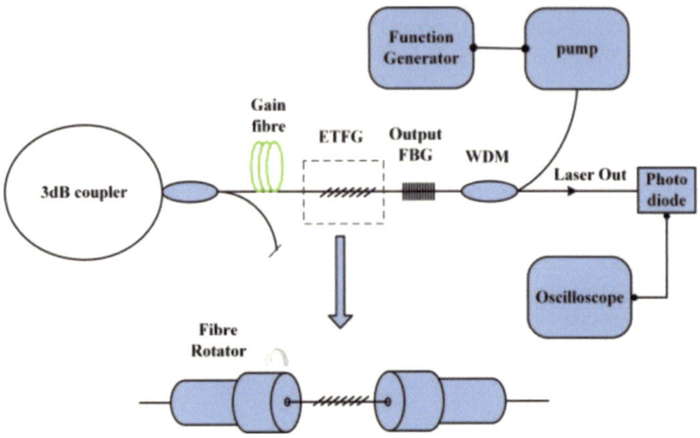

Figure 19. Schematic diagram of the ETFG based fibre laser twist sensor system. (reprint from REF[15] with permission from SPIE).

In the experiment, a segment of laser cavity fibre with ETFG was fixed by a clamp on one side and the other side was mounted on a fibre rotator, as shown in Figure 19a. In order to eliminate the noise induced from other effects such as axial strain and bending, the grating fibre was under small tension to maintain it straight. In the twist sensing experiment, the lower and upper modulation levels of the pump power were first set at 14.6 mW and 43.3 mW, respectively. To perform the measurement, the grating fibre was subjected to twist from 0 to 150° with an elecvation of 10° in both clockwise and anti-clockwise directions. We have measured the built-up time for each twist angle and the normalised results are plotted in Figure 20a. Figure 20a shows that at the initial position, i.e. under 0° twist, the ETFG induced polarisation loss to the laser cavity is at its maximum, so the largest built-up time is expected at this point. Therefore, when the grating fibre is subjected to twist in either clockwise or anti-clockwise direction, the induced polarisation loss through the ETFG to the laser cavity decreases and the built-up time hence reduces accordingly with increasing twist. We have repeated the twist experiment at a different lower pump modulation level of 24 mW. We observed a decrease for the overall torsion sensitivity, as the lower trace shown in Figure 20a. This is because a higher pump power could provide higher gain for the laser system therefore shorter build up time is expected. To work out the sensitivity of the twist sensor, we re-plot the results for twist applied in clockwise and anti-clockwise direction separately in Figure 20b and Figure 20c. Within the dynamic range of ±140°, the sensor shows a quasi-linear response indicating a torsion sensitivity of ~412 μs/(rad/m). The resolution of the oscilloscope used in the experiment is 5 ns, this gives an estimated sensor resolution of ~1.25×10^{-5} rad/m. Furthermore, if the sensor is set at a predefined twist angle, i.e. at ±80° as indicated in Figure 20b and Figure 20c, one is able to identify the twist direction.

http://dx.doi.org/10.5772/57146

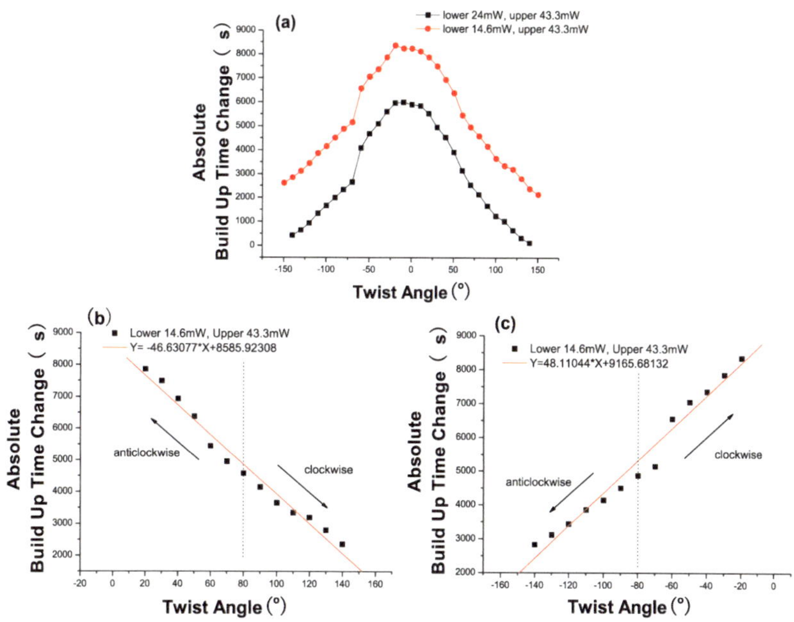

Figure 20. (a) Laser oscillation built-up time against twist angle for twist applied to clockwise and anti-clockwise directions for two different low modulation pump power levels, and re-plotted separately for (b) clockwise and (c) anti-clockwise direction to show the capability of identifying twist direction. (reprint from REF[15] with permission from SPIE).

One may notice that when the torsion angle varies from -140° to 0°, there is a very obvious jump for the built-up time change from -80° to -60°. This could be attributed to the experimental error from rotating the grating in the laser cavity, as there is no such jump for rotation angle from 0° to +140°. To increase the sensitivity of the sensor, it is possible to further decrease the lower modulation pump level.

7. ETFG based loading sensing

Take a standard single mode optical fibre with cylindrical geometry into consideration, when the transverse force is applied to the y axis as shown in Figure 21, for a given compressive force *F*, the stresses in *x* and *y* directions can be expressed as $\sigma_x = \frac{2 \cdot F}{\pi \cdot D \cdot L}$ and $\sigma_y = -\frac{6 \cdot F}{\pi \cdot D \cdot L}$ where *D* is the diameter of the fibre, *L* is the length of the stressed area and *F* is the force applied on the fibre. It is noted that δ_x is tensile stress which is positive while δ_y is compressive stress which is negative therefore $(\delta_x - \delta_y) > 0$ is always true. The photoelastic effect induced refractive index change in the fibre core area can be given by [22]:

$$\Delta n = n_x - n_y = \left(n_{x0} - n_{y0}\right) + \left(C_1 + C_2\right) \cdot \left(\delta_x - \delta_y\right) \tag{1}$$

Where n_{x0} and n_{y0} are the effective refractive indices of the fibre without stress. C_1 and C_2 are the stress-optical coefficient, the relationship (C1-C2)>0 is always true for silica fibre [23]. If the transverse load is applied in the slow-axis of an ETFG, we will have $n_{x0}=n_f$ and $n_{y0}=n_s$ where n_f and n_s are the refractive indices for the predefined fast- and slow-axis of the ETFG. Hence, the first term in equation (1) will be negative which will therefore reduce the birefringence Δn. Under this situation, we anticipate the light coupling to the two orthogonal polarised modes is apt to be affected by the external loading. On the contrary, if the transverse load is applied in the fast-axis direction, we will have $n_{x0}=n_s$ and $n_{y0}=n_f$. Therefore, we will have a positive value in the first term of equation (1) which will increase the birefringence Δn. In this scenario, the ETFG is capable of preventing light from coupling to the two orthogonal polarised modes.

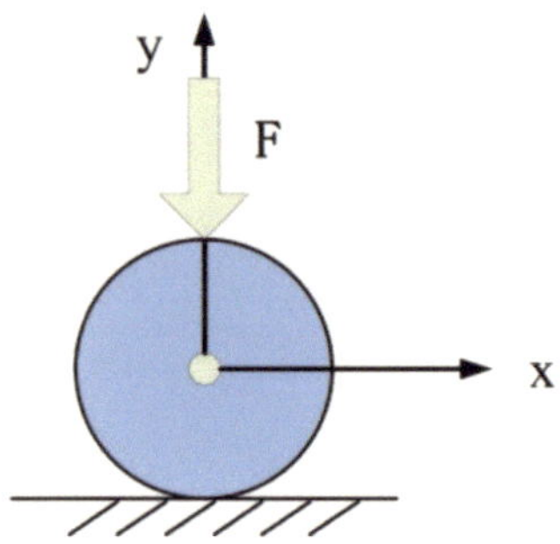

Figure 21. The cross section of a fibre in an assigned *x–y* coordinate system with transverse load applied along the *y*-axis.

The schematic experimental setup of the ETFG loading experiment is shown in Figure 22a. In the experiment, an ETFG with 12 mm length and a support fibre were sandwiched between the flat surface of a buffer plate and a base plate. The loading was applied on top of the buffer plate with a nominal loading length of ~ 32 mm. A BBS was polarised through a commercial polariser, the PC was employed to choose the desired polarisation state, here we choose, the fast-axis mode of the ETFG. The spectral evolution was recorded through an OSA. The applied transverse load on the slow-axis of the ETFG was from 0 to 2600 grams with an increment of 200 grams as shown in Figure 22b. The spectral evolution is plotted in Figure 23. It can be seen clearly from Figure 23 that with the increase of the loading weight, the intensity of the fast-axis mode is elevating while that of the slow-axis mode is decreasing. While we found when the loading is applied on the fast axis of the ETFG as shown in Figure 22c, no spectral evolution can be observed. This unique property could potentially serve as a vectorial loading sensor which is capable of not only measuring the amplitude of the loading but also identifying the direction of the loading.

Similar to the twist sensor, the loading sensor can also be optimised into a low cost sensor using a SWL and optical power detector. There also exists a linear range from 9 to 44 kg/m for the loading responsivity which is approximately 2.04 μW/(kg/m). Therefore, by using a standard detector with 1nW resolution, one could possibly reach a loading sensitivity of 0.0016 grams.

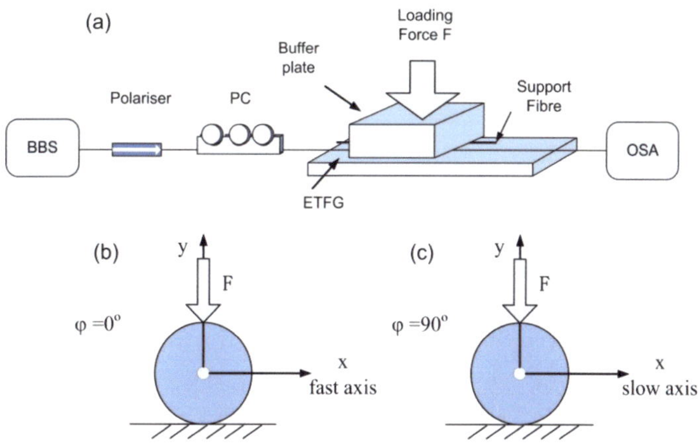

Figure 22. (a) Schematic experimental setup of the transverse loading experiment using an ETFG. (b) and (c) The cross section of the TFBG with transverse load applied along slow- and fast-axes; φ is the angle between the fast-axis of the ETFG and the *x* direction.

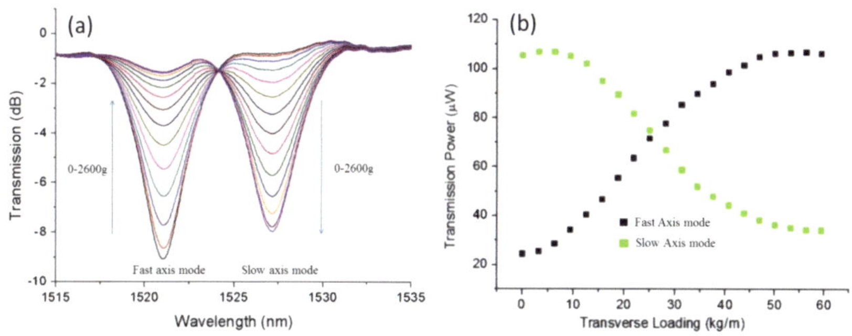

Figure 23. (a)Transmission spectral evolution of the ETFG with transverse load from 0 to 2600g applied on the slow-axis of the ETFG and the support fibre. (b) Power evolution of the ETFG loading sensor using SWL and power meter.

8. ETFG based refractive index sensing

Due to the nature of forward propagating cladding modes, the ETFG is intrinsically sensitive to the external refractive index variation which is similar to an LPG. To characterise the RI responsivity, the ETFG was mounted in a V-grooved aluminium plate, this is to guarantee the measurement being free from other effects such as strain and bend. The refractive index gel used to study the RI response of the ETFG is certified commercial gels from Cargille Lab. The optical spectrum was recorded through a BBS and an OSA to study the spectral evolution of RI response. In Figure 24, it clearly shows the two paired modes around 1560 nm and 1610 nm in response to the RI gels with index change from 1.32 to 1.38 with an increment of 0.01. An obvious trend is the degeneration of the dual loss peaks with increasing RI value. This can be explained by the reduction of polarisation mode dispersion (PMD) of the cladding modes resulting from the decrease of refractive index difference between the fibre cladding and outer medium. Figure 25a and Figure 25b shows the RI responses of the fast- and slow-axis mode for 1560 nm and 1610 nm band, respectively. It can be clearly seen that the wavelength shift for the two bands is around ~20 nm when the RI increased from 1.32 to 1.38. Additionally, the wavelength shift for 1610 nm band is observed to be slightly larger than that of the 1560 nm band. It is also obvious to see that the ETFG has stronger RI response for the RI value around 1.3 compared to conventional LPGs [20] which does not respond at this RI range. This indicates that ETFG could be an ideal candidate as bio/chemical sensor for water based solutions. Despite the RI response within the desired RI range is not linear, the RI sensitivity of the two peaks is estimated to be ~320 nm/RIU in ~1.33 refractive index region, which is much higher than that of reported typical LPGs [20]. Further increase the RI may result in RI match between the cladding and the index gel. This will remove the boundary condition for the survival of cladding modes, therefore only radiation mode will appear without any loss band structure in the optical spectrum.

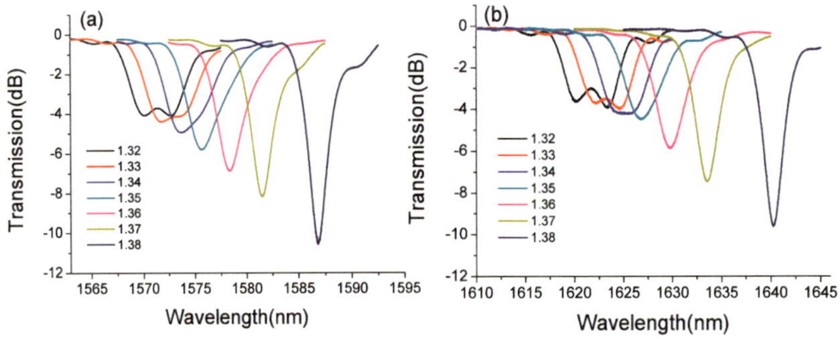

Figure 24. RI response of a mode pair when subject into index gel from RI=1.32 to 1.38 with an increment of 0.01 for (a) 1560 nm band and (b)1610 nm band (Reprinted from Optics Communication Vol.305 pp271-275, Copyright (2013) with permission from Elsevier).

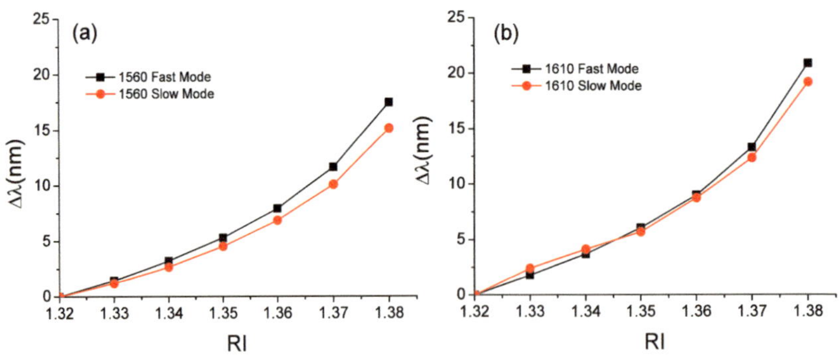

Figure 25. RI response of fast- and slow- axis modes around (a) 1560 nm and (b) 1610 nm when the ETFG was subjected to index gels with SRI change from 1.32 to 1.38 with an increment of 0.01 (Reprinted from Optics Communication Vol.305 pp271-275, Copyright (2013) with permission from Elsevier).

9. Liquid level sensor

Based on the RI response of the ETFG, we notice that when the ETFG is fully immersed in water, the paired peaks shifted to the longer wavelength side, as shown in Figure 26. It is this water induced wavelength shift enables ETFG to be an effective liquid level sensor. If part of an ETFG is immersed in the water, it behaves as two individual gratings with two types of surrounding media. Hence, when we just observe one paired loss peaks, we will see that a paired loss peaks generated by the air-surrounded grating section with a broad peak on longer wavelength side generated by the water-surround grating sections, as illustrated in Figure 26. This behaviour exhibits similarity to LPG and etched FBG based level sensors, and further proves that the effective index of the propagating cladding or core mode is defined by the RI of the surrounding medium covering the waveguide. Owing to the PMD nature of the ETFG cladding modes, we have excited fast- and slow-axis mode individually by polarised light to evaluate the ETFG spectral response to water level change.

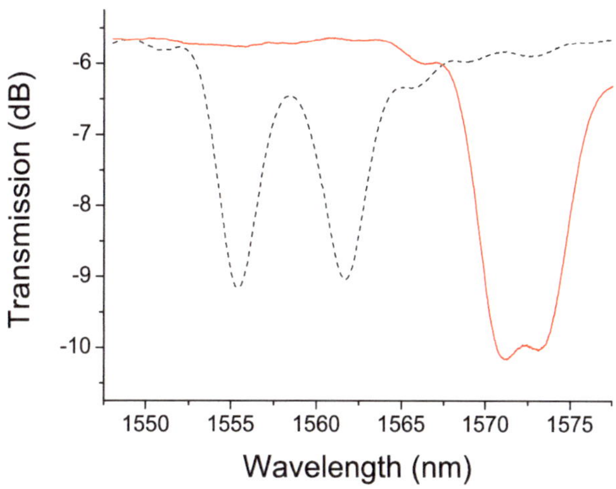

Figure 26. The transmission spectra of a 78°-ETFG when it is in air (black solid line) and fully immersed in water (red dotted line) (Reprinted from Optics Communication Vol.305 pp271-275, Copyright (2013) with permission from Elsevier).

In the water level sensing experiment, a 12 mm long grating was attached to a plastic tube and then immersed into the water. One part of the tube surface was removed to house the grating fibre which is to ensure the grating is in full contact with water while also eliminating the grating from bend. A precision translational stage was used to hold the beaker so that water level on the submerged ETFG can be accurately controlled. Hence during the whole experiment, the fibre movement can be eliminated which stops the polarisation induced uncertainty. Figure 27a describes the schematic configuration of the level sensing experiment. In Figure 27b, it shows the transmission spectra when the slow-axis mode around 1560 nm band was excited at three different liquid levels: the left peak (red solid line) indicates that the grating fully exposed to air, the middle two peaks (blue dashed line) appear when half of the grating is immersed in water, the right peak (black dotted line) illustrates that the grating is completely immersed in water. The spectral evolution of the slow-axis mode around 1560 nm when the ETFG was surrounded completely from by air to water is illustrated in Figure 27c. It is clear to see that when the water level increases, the loss of the air-surrounded peak decreases whereas the loss of the water-surrounded peak elevates. Figure 27d shows that the fast-axis mode around 1560 nm band exhibited a similar spectral change,

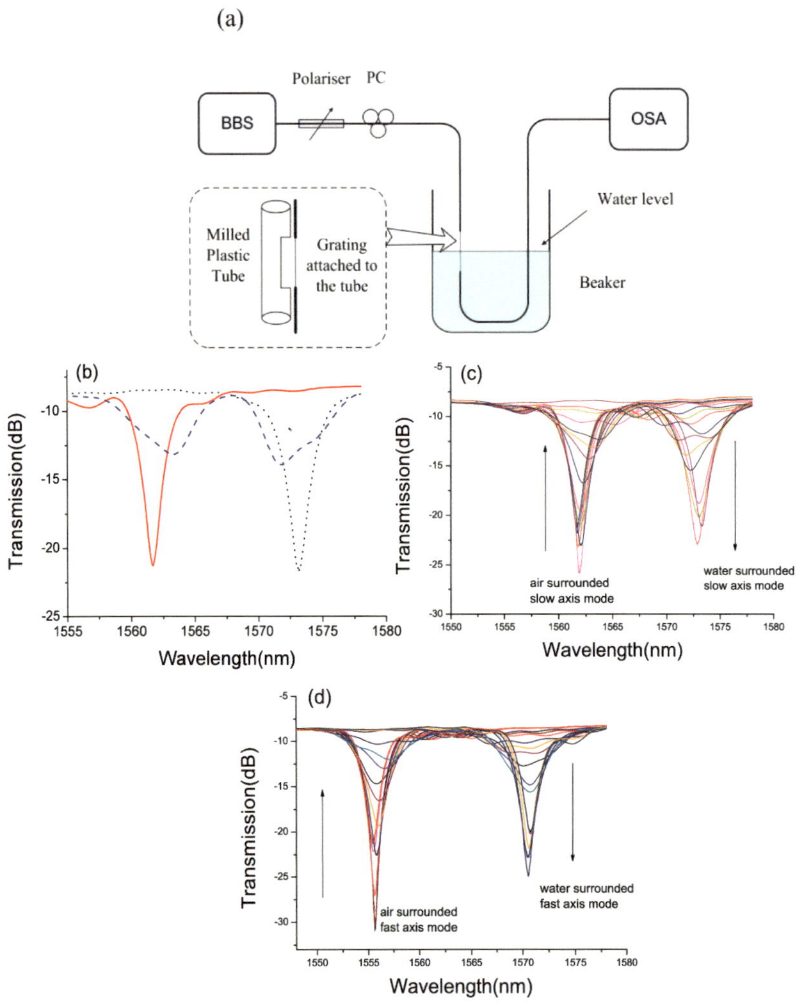

Figure 27. (a) Schematic diagram of the experimental setup for liquid level sensing; (b) Transmission spectra of the ETFG for slow-axis mode coupling around 1560 nm: the left-side (red solid line), right-side (black dotted line) single peak and the middle dual peaks (blue dashed line) corresponding to the grating surrounded by air, water and half way in water. The air- and water-surrounded peaks evolving with increasing water level for (c) slow-axis mode and (d) fast-axis mode (Reprinted from Optics Communication Vol.305 pp271-275, Copyright (2013) with permission from Elsevier).

The transmission against submerging length of the ETFG in water for the fast- and slow-axis mode are depicted in Figure 28a and Figure 28b respectively.

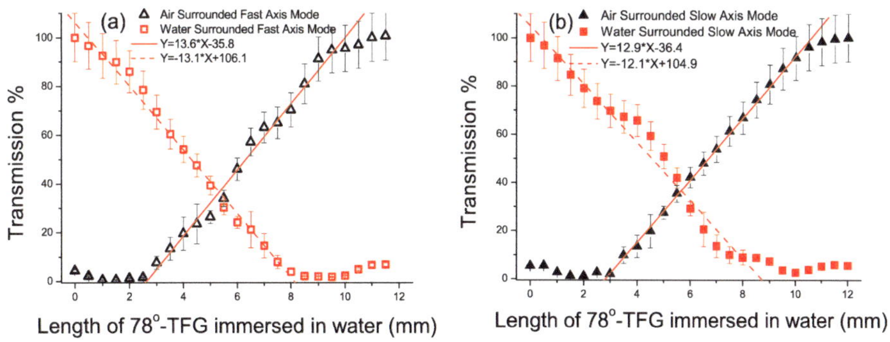

Figure 28. Transmission change measured for the length of the 78-TFG in water for the grating sections surrounded by water (■,□) and by air (▲,△) for (a) fast-axis mode and (b) slow-axis mode (Reprinted from Optics Communication Vol.305 pp271-275, Copyright (2013) with permission from Elsevier).

In Figure 28, it can be seen that by observing the transmission of the fast-axis or slow-axis mode and the related water surrounded peak of the ETFG, we can determine the water level. Additionally, one need to note that the initial 3 mm for the air-surrounded modes is not sensitive to the water level variation. The grating has a linear response to the water level change from 3 mm to 10 mm and becomes insensitive from 10 mm to 12 mm. While for the water-surrounded modes, the grating response has a linear relationship to water level from 0 mm to 8 mm followed by an insensitive range from 8 mm to 12 mm. It is still not clear that why the sensing range is different between the air- and water-surrounded modes. We further linearly fitted the results to characterise the sensitivity of the level sensor.

In Figure 28a, it can be seen the sensitivity for the air-surrounded fast-axis modes is quite similar to that of the water-surrounded one which is ~ 13 %/mm despite the sensing range is different for these two modes. The air- and water-surrounded slow-axis mode show a sensitivity of ~ 12 %/mm which is shown in Figure 28b. This is more than twice of the sensitivity of the LPG based liquid level sensor with a reported value of 4.8 %/mm [24]. We can therefore conclude that both fast- and slow-axis mode of the ETFG is capable of water level sensing which offers similar sensitivity and measurement range. The manufacture quoted accuracy of 7 % (0.3 dB) of the OSA can be regarded as the main factor of measurement uncertainties in the experiment. Also, the environmental effect could be taken into account because the measurement is polarisation sensitive. The uncertainty may increase up to ~ 35 % if the transmission of the water-surrounded mode is getting lower. This is mainly due to reaching the sensitivity limit of the OSA when the transmission of the peaks become smaller. A stronger light source and low loss measurement kit could potentially increase the sensitivity and eliminate these errors. The sensor may also be developed as a low cost solution using single wavelength laser and a power meter.

10. Conclusion

In summary, ETFGs have been demonstrated as effective RI, strain, twist, loading and liquid level sensor. The ETFGs have shown a slightly low thermal responsivity and slightly higher strain responsivity than the standard FBGs. The strain response of the ETFGs is also similar to a short period LPG as shown a negative wavelength shift against strain. Because their unique polarisation mode splitting property, the ETFGs have exhibited vectorial sensing functions as loading and twisting sensors. This allows the ETFG can not only measure the amplitude of the loading and twist but also determine the direction of the measurands. We have described that as a strain or twist sensor, the signal can be demodulated using low cost method with a single wavelength laser and a power detector. Furthermore, we have demonstrated that by incorporating the ETFG in a linear cavity fibre laser, a fibre laser based strain or twist sensor using a time domain signal demodulation method can be realised showing high signal-to-noise ratio for the sensing. The ETFG has also shown strong responsivity to external surrounding medium. Compared to an LPG, the ETFG is sensitive in the RI range ~ 1.33 which allows it to perform efficient sensing for aqueous based solution. Based on this, we have also demonstrated an ETFG based water level sensor with higher sensitivity than an LPG based one.

In the future, we will expect to see more sensor applications based on ETFGs, such as bending sensor and biophotonic application. With proper coating, the ETFGs could function as surface plasmon resonance based ultrahigh sensitivity sensors. So far the theoretical understanding of ETFGs has not been fully realised, thus a systematic study on the theory of the ETFG is expected with which we could find more applications with such unique grating structures.

Author details

Chengbo Mou*, Zhijun Yan, Kaiming Zhou and Lin Zhang

*Address all correspondence to: moucl@aston.ac.uk

Aston Institute of Photonic Technologies, School of Engineering and Applied Science, Aston University, Aston Triangle, Birmingham, UK

References

[1] Hill KO, Fujii Y, Johnson DC, Kawasaki BS. Photosensitivity in Optical Fiber Waveguides - Application to Reflection Filter Fabrication. Appl Phys Lett. 1978;32(10): 647-9.

[2] Meltz G, Morey WW, Glenn WH. Formation of Bragg Gratings in Optical Fibers by a Transverse Holographic Method. Opt Lett. 1989;14(15):823-5.

[3] Rao YJ. In-fibre Bragg grating sensors. Meas Sci Technol. 1997;8(4):355-75.

[4] Anderson DZ, Mizrahi V, Erdogan T, White AE. Production of in-Fiber Gratings Using a Diffractive Optical-Element. Electron Lett. 1993;29(6):566-8.

[5] Bhatia V, Vengsarkar AM. Optical fiber long-period grating sensors. Opt Lett. 1996;21(9):692-4.

[6] Peupelmann J, Krause E, Bandemer A, Schaffer C. Fibre-polarimeter based on grating taps. Electron Lett. 2002;38(21):1248-50.

[7] Miao YP, Liu B, Zhang H, Li Y, Zhou HB, Sun H, et al. Relative Humidity Sensor Based on Tilted Fiber Bragg Grating With Polyvinyl Alcohol Coating. Ieee Photonic Tech L. 2009;21(7):441-3.

[8] Caucheteur C, Chen C, Albert J, Mégret P. Use of weakly titled fiber Bragg gratings for strain sensing purposes. Proc of IEEE/LEOS Benelux Chapter, Eindhoven. 2006:61-4.

[9] Guo T, Tam HY, Krug PA, Albert J. Reflective tilted fiber Bragg grating refractometer based on strong cladding to core recoupling. Opt Express. 2009;17(7):5736-42.

[10] Shao LY, Xiong LY, Chen CK, Laronche A, Albert J. Directional Bend Sensor Based on Re-Grown Tilted Fiber Bragg Grating. J Lightwave Technol. 2010;28(18):2681-7.

[11] Shevchenko YY, Albert J. Plasmon resonances in gold-coated tilted fiber Bragg gratings. Opt Lett. 2007;32(3):211-3.

[12] Mihailov SJ, Walker RB, Stocki TJ, Johnson DC. Fabrication of tilted fibre-grating polarisation-dependent loss equaliser. Electron Lett. 2001;37(5):284-6.

[13] Zhou KM, Simpson G, Chen XF, Zhang L, Bennion I. High extinction ratio in-fiber polarizers based on 45 degrees tilted fiber Bragg gratings. Opt Lett. 2005;30(11):1285-7.

[14] Zhou KM, Zhang L, Chen XF, Bennion I. Optic sensors of high refractive-index responsivity and low thermal cross sensitivity that use fiber Bragg gratings of > 80 degrees tilted structures. Opt Lett. 2006;31(9):1193-5.

[15] Mou C, Zhou K, Suo R, Zhang L, Bennion I, editors. Fibre laser torsion sensor system using an excessively tilted fibre grating and low-cost time domain demodulation. 20th International Conference on Optical Fibre Sensors; 2009 Oct 5-9; Edinburgh,UK2009.

[16] Chen X, Zhou K, Zhang L, Bennion I. In-fiber twist sensor based on a fiber Bragg grating with 81 degrees tilted structure. Ieee Photonic Tech L. 2006;18(21-24):2596-8.

[17] Suo R, Chen XF, Zhou KM, Zhang L, Bennion I. In-fibre directional transverse loading sensor based on excessively tilted fibre Bragg gratings. Meas Sci Technol. 2009;20(3):034015.

[18] Mou C, Zhou K, Yan Z, Fu H, Zhang L. Liquid level sensor based on an excessively tilted fibre grating. Opt Commun. 2013;305:271-5.

[19] Zhou KM, Zhang L, Chen XF, Bennion I. Low thermal sensitivity grating devices based on Ex-45 degrees tilting structure capable of forward-propagating cladding modes coupling. J Lightwave Technol. 2006;24(12):5087-94.

[20] James SW, Tatam RP. Optical fibre long-period grating sensors: characteristics and application. Measurement Science and Technology. 2003;14:R49-R61.

[21] Yang XF, Luo SJ, Chen ZH, Ng JH, Lu C. Fiber Bragg grating strain sensor based on fiber laser. Opt Commun. 2007;271(1):203-6.

[22] Gafsi R, El-Sherif MA. Analysis of induced-birefringence effects on fiber Bragg gratings. Opt Fiber Technol. 2000;6(3):299-323.

[23] Imoto N, Yoshizawa N, Sakai JI, Tsuchiya H. Birefringence in Single-Mode Optical Fiber Due to Elliptical Core Deformation and Stress Anisotropy. Ieee J Quantum Elect. 1980;16(11):1267-71.

[24] Khaliq S, James SW, Tatam RP. Fiber-optic liquid-level sensor using a long-period grating. Opt Lett. 2001;26(16):1224-6.

Chapter 3

Optical Sensors Based on Mesoporous Polymers

Ruslan Davletbaev, Alsu Akhmetshina,
Askhat Gumerov and Ilsiya Davletbaeva

Additional information is available at the end of the chapter

http://dx.doi.org/10.5772/57383

1. Introduction

Porous materials are widely used in such industries as chemical, food industry, petrochemistry, medicine, and environmental protection. Nanoporous materials (NM), characterized by a pore size of 1 to 100 nm, are a great alternative to non-porous materials due to the presence of a number of unique properties. Among NM are microporous materials (e.g. zeolites) and mesoporous materials (e.g., porous polymers, aluminum or silicates). According to IUPAC nomenclature, the definition of "microporous" corresponds to the pore size of 2 nm, the definition of "mesoporous" corresponds to 2-50 nm [1]. By chemical composition NM are divided into aluminum silicates, metals, oxides, silicates, consisting only of carbon and organic polymers. These materials are combined in a high surface area and porosity. NM are used in various fields of chemistry and technology, depending on their chemical composition, pore size and distribution, porosity value.

Promising areas in which mesoporous materials can be successfully used are transparent optical chemical sensors and test methods for the determination of various substances. Sensors and sensing elements on optically transparent polymeric substrates can be more convenient in some cases of analytical practice, as they allow to observe visually the color change [2]. To determine the low concentrations of elements preliminary sorption concentration and subsequent determination by chemical or physico-chemical methods are used. The optical transparency of the sorbent allows to carry out the analytical reactions on sorbent surface and on their basis to develop sorption-photometric and test methods for the determination of substances.

Complexation reactions of organic reagents and a tested ion are the basis of optical chemical sensors for metal cations, they are accompanied by color change of the reaction system. In absorption-based optical sensors, the molecules of organic chromophores are used as a sensing layer of the substance (receptor), selectively interacting with the analyte [3]. In the case of an luminescence-based optical sensor, the molecules of organic luminophorsare receptors. The latter also became common as active medium in dye lasers.

Indicator papers, indicator tablets, powders, solutions in vials may be used as a substrate. There is a number of substrates, such as cellulose, ion-exchange resins, superfine silicas, polyvinyl chloride membranes, etc. where the complexing reagent can be immobilized. The main requirements to the material of the reagent carrier are optical transparency, high sorption rates, ease of synthesis, processability, inertness to the reactants, stability in acidic and alkaline media, high sensitivity to analytes [4].

The natural cellulose polymer is the most widely used as a matrix [5]. A solid carrier is soaked in the reagent solution and then dried. The process can be one-stage or multistage.

The methods of applying reagents to the polyurethane foams are worked out. These methods are based on pre-plasticization of polyurethane foam tablets and the subsequent treatment by small volume of acetone solution of an analytical reagent [6].

Silica gel is a porous, granular silica form synthetically produced from solutions of sodium silicate or silicon tetrachloride, or substituted chlorosilanes / orthosilicates. The active surface of silica gel with a large surface area is of great importance in the adsorption and ion exchange. The modification of the silica gel surface for the analytical reactions is carried out in two separate ways, viz. the organic functionalization, where the modifier is an organic group; and inorganic functionalization where a group fixed on the surface can be an organometallic composite or metal oxide [7].

One of the interesting areas of research is gelatin cured gel applied to the substrate made of a transparent polymer. Gelatin is a polydisperse mixture of polypeptides prepared by alkaline or acid hydrolysis of collagen. Biopolymer has high hydrophilicity, transparency in the visible spectrum, ability to form gels at any weight ratio of water-gelatin. These properties make the systems with immobilized reagents on the basis of such polymer especially attractive for analytical reactions involving water as a solvent [8].

In recent years modified ion-exchange materials prepared by sorption, or chemical grafting of organic reagents using conventional ion exchangers, have found wide application [9]. Such ion exchangers are used for the selective and group concentration of elements, but the most valuable thing is that owing to them, an appropriate element can be simultaneously concentrated and determined quantitatively.

There are a number of shortcomings for the known substrates which limit the prevalence of these materials as the base of optical chemical sensors. Cellulose is non-transparent, has low resistance to aggressive environments, weak physical and mechanical properties. Polyurethane foams are also non-transparent and characterized by high desorption of the chromophores from the pores of the material. Gelatin gels have low physical and mechanical properties. Ion-exchange sorbents are characterized by the complexity of synthesis and low kinetic characteristics of sorption.

Therefore, the search for new optically transparent materials with developed specific surface, chemical resistance and high physical and mechanical properties is an urgent task of polymer chemistry and materials science

2. Fundamentals of synthesis of mesoporous polymers based on aromatic isocyanates

Isocyanates are able to enter into chemical reactions leading to the formation of polymers with different structure. In most cases for the isocyanates, the reactions of nucleophilic addition of compounds containing mobile hydrogen atoms are typical. The reactions of isocyanates with diols and diamines are the most important from a practical point of view. In the presence of catalysts (tertiary amines, alkoxides and carboxylates of quaternary ammonium base, etc.) isocyanates go through dimerization and trimerization to form uretidinedions and isocyanurates [10]. The possibility of catalytic homopolymerization of isocyanates by anionic mechanism is also known. As a rule, polyisocyanates of amide nature are polyaddition products. However, in the literature [11] there is information about the formation of polyisocyanate links of acetal nature (O-polyisocyanates). The possibility of transformations in different directions is due to the ambident nature of the anion at the end of the growing chain (Figure 1). These works give the information about obtaining the polyisocyanate links of acetal structure by copolymerization of ethylene oxide and aromatic isocyanates using IR spectroscopy and chemical degradation of the polymer. Typical for these polymers, N = C bond appears in the IR spectra in the region of 1670-1680 cm^{-1}.

In [12,13] it was established that the open chain analogs of crown ethers, which are block copolymers of ethylene oxide and propylene oxide containing terminal and potassium alcoholate groups, are effective initiators of opening the isocyanate groups along the thermodynamically more stable carbonyl group. It was assumed that the capture of the metal cation by polyester fragment acting as a linear podand promotes the preferential localization of the negative charge on the oxygen atom of the growing chain in anionic polymerization. It was also shown that such polyaddition occurs only when 2,4-tolylene diisocyanate is used involving isocyanate groups of more active para-position to the reaction process.

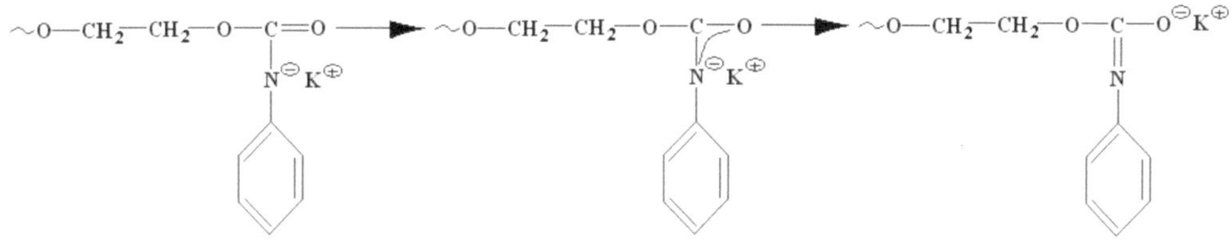

Figure 1. Ambident nature of the anionic center in the reaction of 2,4-toluene diisocyanate with alcoholates

Open chain analogs of crown ethers, which are block copolymers of ethylene oxide and propylene oxide containing terminal potassium alcoholate groups turned out to be effective initiators of isocyanate groups opening by anionic mechanism. Previously, it was found that the opening of the isocyanate groups along the N = C bond led to cyclization of polyaddition products with subsequent formation of polyisocyanurates. However, it turned out that if the macroinitiators were open-chain analogues of crown ethers there could be created favorable conditions for the opening of the isocyanate groups along the carbonyl component and the formation of the polyisocyanate structures of the acetal nature. Earlier in [14], it was shown that the polymers obtained in this way can reach the free volume up to 20% due to the formation of mesopores. Until now, there was uncertainty about the causes of the formation of transition pores in the volume of the polymer, the percentage of O-polyisocyanate blocks and how to manage the process in terms of creating the most favorable conditions for the formation of mesoporous polymers. According to the chemical structure of O-polyisocyanate links, their macroblocked structures are able to exhibit intense intermacromolecular interactions leading to their segregation.

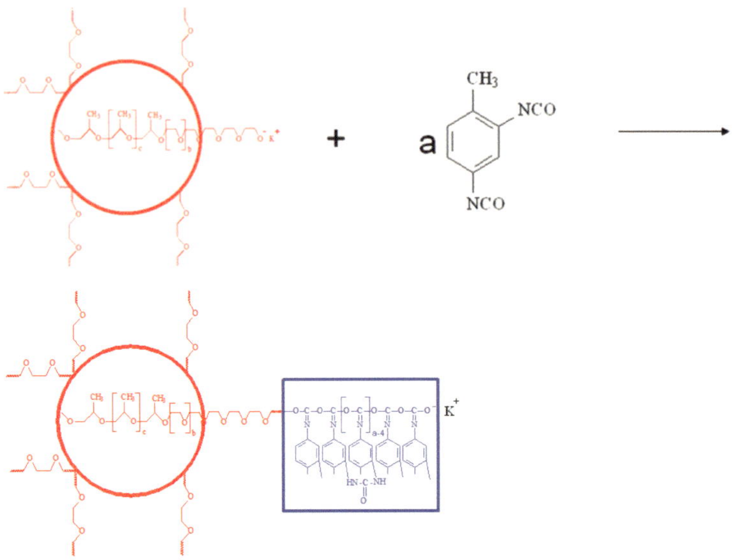

Polymerization starts from a macromolecular coil representing the block copolymers of ethylene oxide and propylene oxide. Terminal reactive O-polyisocyanate units exhibit the ability to stabilize due to their interaction with forming structural fragments of crown ethers in the coil of macroinitiators.

In turn, the coil of macroinitiator is quite large, larger than 50 nm. The result is that the coil of macroinitiator contains a great number of macromolecules of open-chain analogs of crown ethers and is able to engender the growth of O-polyisocyanates' units as well as to stabilize the reactive terminal units.

http://dx.doi.org/10.5772/57383

In [15], it was shown that the involvement of the isocyanate groups in the *ortho*-position into the reaction with latent water with the following formation of urea groups contributed to stabilization of O-polyisocyanate blocks.

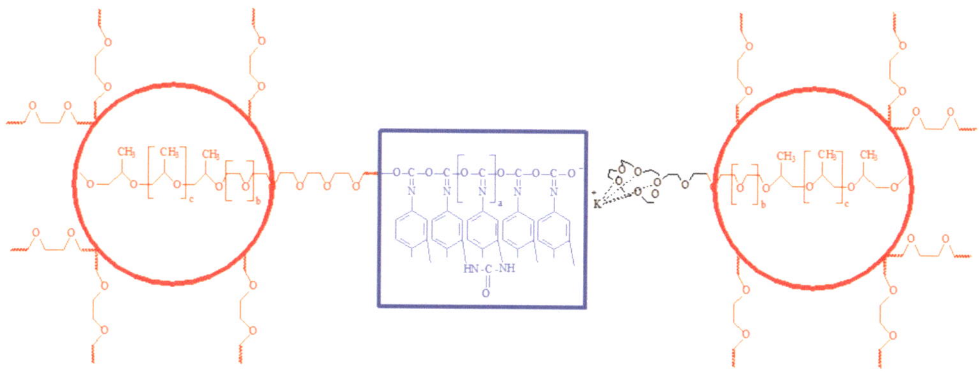

In this case the initiation and stabilization acquired rigorous multi-dimensional geometry in the space, the result of which was the cell structure with voids similar to honeycombs (Figure 2).

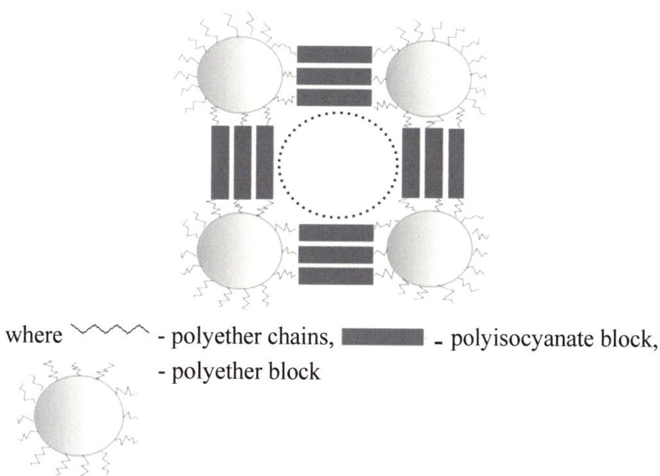

Figure 2. Scheme of the pores formation in the polymer based on anionic macroinitiator and 2,4-toluene diisocyanate

Evaluation of the sorption capacity of the polymers was carried out by water absorption. The adsorption isotherm of water vapor for the polymer sample was S-shaped, which is typical for polymers with transition pores (Figure 3). The specific surface area of sorbents was calculated by the equation proposed by Brunauer, Emmett and Teller and was 75 m^2/g. According to [16], the specific surface of transitional pores (mesopores) is in the region of 20-200 m^2/g.

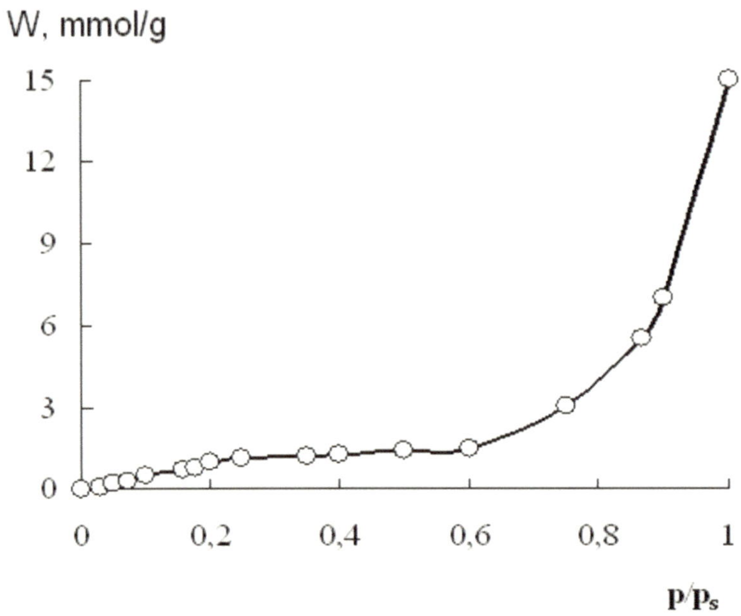

Figure 3. Water sorbtion curve for mesoporous polymer

Additional information about the processes of geometry of supramolecular structures in these polymers was obtained by using atomic force microscopy (Figure 4).

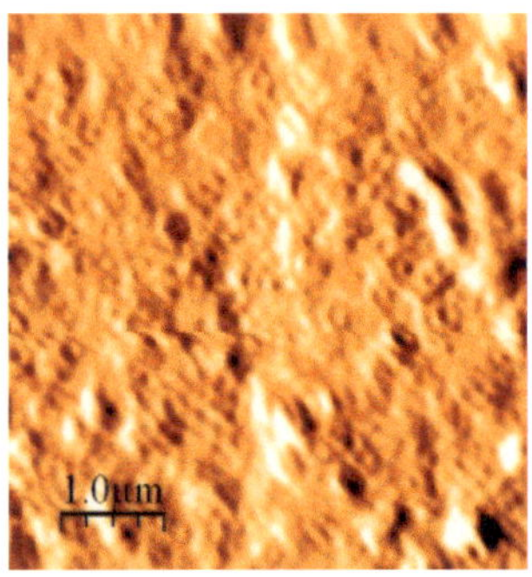

Figure 4. AFM image of a mesoporous polymer

3. Features of the organic chromophores immobilization in mesoporous polymers

Reactants, slightly soluble in water, are preferred, since the test-forms are more stable during storage and they are weakly leached from the test matrix during the contact with the test liquid.

When slightly water-soluble reactants are immobilized on the carriers, their solutions in organic solvents are used or the reagents are applied in the form of fine powder. To increase the binding strength of the agent with the carrier the chemical bonds are formed between them (chemical immobilization). For immobilization by covalent bonding (chemical immobilization) cellulose, polymers sorbents and silica gels are used as carriers. However, the "physical" fixing is usually much simpler, so it is quite widespread.

The presence of voids in the mesoporous polymer was the reason to immobilize organic polymer-supported reagents. The study of sorption processes of luminophores in polymers is of interest in connection with the possibility of obtaining laser media and sensors for photometric and luminescent determination of metal cations.

The study of the sorption properties of the polymeric material with respect to organic luminophores was carried out by electronic spectroscopy (Figure 5).

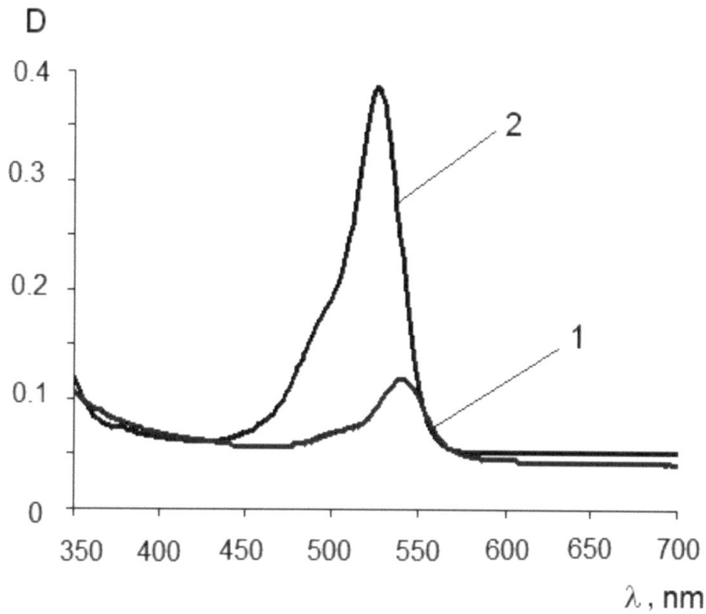

Figure 5. Electronic spectra of the solution of rhodamine 6G (1) in ethanol and rhodamine 6G, immobilized in the mesoporous polymer (thickness of the cuvette for the solution and the polymer sample was 0.1 cm)

As the luminophore rhodamine 6G (R6G) was selected and used to determine gold cations. A solution of R6G of the given concentration was obtained and the maximum value of the optical density was determined. According to the Bouguer-Lambert law the molar absorption coefficient for the solution of rhodamine 6G was calculated. Rhodamine 6G was sorbed on the polymer. The value of the optical density of the polymer sample doped with R6G at $\lambda_{max} = 530$ nm was measured. The calculation of dye concentration in the polymer was performed. It was established that electron spectra of organic agents on polymer carriers did not have substantial changes in comparison with the spectrum of their solutions. It was established that adsorption of organic phosphor 6G rhodamine is accompanied by its concentration of polymer matrix. The dye concentration in the polymer was 3 times more than in the initial solution. This fact points to the flow of intense adsorption of rhodamine 6G in the voids of the polymer.

4. Mesoporous polymers as laser active media

Dye lasers are used for spectroscopic studies to improve the sensitivity, spectral and temporal resolution by several orders of magnitude compared to traditional methods of spectroscopy. They can also be used where high energy of laser radiation is not needed. Typically, in dye lasers the solutions of dyes (solvents - water, alcohols, benzene derivatives, etc.), rarely dyes activated polymeric materials - polymethyl methacrylate, epoxy resin, polyurethane, etc. are used which are called polymeric laser-active media. However, these media have a number of drawbacks that make them difficult to use. For example, the polymer and dye undergo relatively rapid photodegradation, so the active medium often has to be changed.

Nowadays, thanks to the efforts of chemists and physicists, solid-state active media for tunable lasers are made with parameters as good as solutions' parameters. In recent years the interest in the emission of organic molecules in the thin films is growing due to the possibility of using them as the base for photoexcitating microlasers and the materials for OLEDs. Polymers have great advantages over other materials. They show high optical uniformity, good compatibility with organic dyes and at the same time they are cheap and manufacturable. The latter facilitates the miniaturization and embeddability in optical systems.

Laser properties of mesoporous and nonporous polymers doped with rhodamine 6G dye were investigated for comparison. Because of specific samples forms it wasn't possible to measure laser efficiency directly. To estimate the laser efficiency, operating life-time amplified spontaneous emission (ASE) under transverse pump by the second harmonic of Q-switched Nd:YAG laser (pulsewidth was 12 ns, pulse repetition rate was 10 Hz) was measured. The pumping region had a form of a stripe with 27 μm width and length close to the sample length. Maximum intensity at the beam waist at the sample was 25 MW/cm^2. The ASE was observed from two opposite samples cuts inside the cones with axis parallel to pump region in such setup configuration. The intensity of the ASE was measured from one side of the investigated sample with piroelectric energy sensor Ophir PE-9 (ASE was focused by spherical lens). Simultaneously from the other side of the sample ASE spectra were measured with wide-range spectrometer S100. At the pump intensity 25 MW/cm^2 the spectral half-width of the ASE didn't exceed 5 nm for all investigated samples (Figure 6).

Photostability of Rhodamine 6G in the mesoporous polymer at low pump energies (12.5 MW/cm^2) amounted to 115000 pulses, at high energies (25 MW/cm^2) - 60000 pulses. The energy of stimulated emission during irradiation reached maximum and then began to fall compared to the original value. The most widespread material for the solid-state polymer dye lasers is nonporous polymethylmethacrylate (PMMA). Therefore, the PMMA with R6G was chosen for comparison. Halving ASE energy of the PMMA doped with R6G was observed at around 200

http://dx.doi.org/10.5772/57383

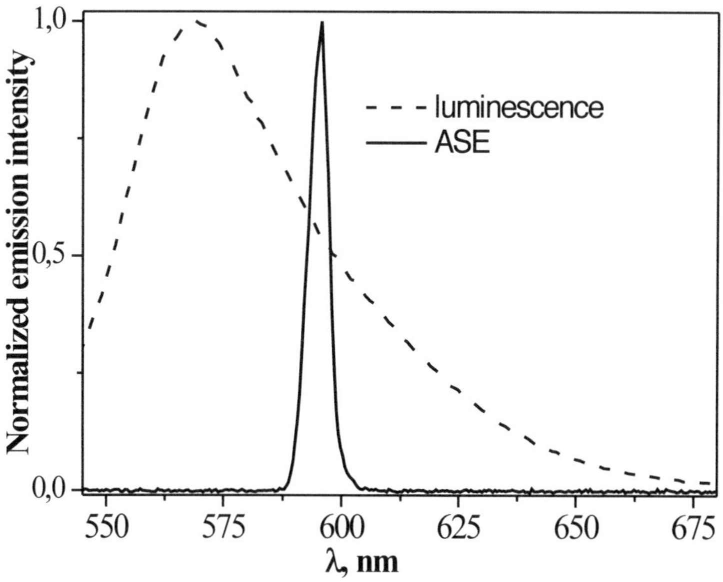

Figure 6. Luminescence (dash line) and ASE (solid line) spectra of the mesoporous polymer doped with rhodamine 6G

pulses, that is, about 175 times smaller than for the mesoporous polymer. PMMA demonstrates exponential ASE decay (Figure 7).

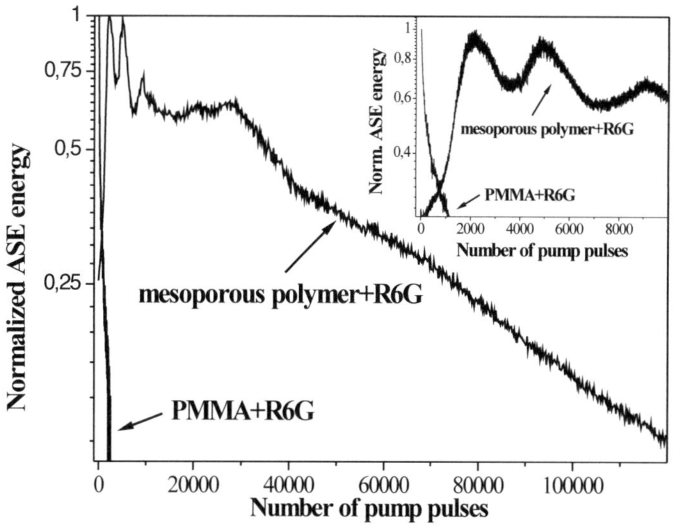

Figure 7. Dependences of the normalized ASE energy on the number of excitation pulses of mesoporous polymer and PMMA doped with R6G (pump intensity was 25 MW/cm^2). The inset shows the details of the dependences at the beginning of the pump

Thus, the processes of generation of laser radiation and photochemical aging of organic luminophors in mesoporous and non-porous polymers were studied. It was shown that in the mesoporous polymers organic luminophors were able to generate laser radiation and high radiation resistance.

5. Mesoporous polymers as basis for optical chemical sensors

Organic chromophores react selectively with ions of many metals forming chelate complexes which are intensely colored. The reactions of complex forming organic chromophores and the ion being identified accompanied by color change of the reaction system are the foundation of chemical test methods on metal cations.

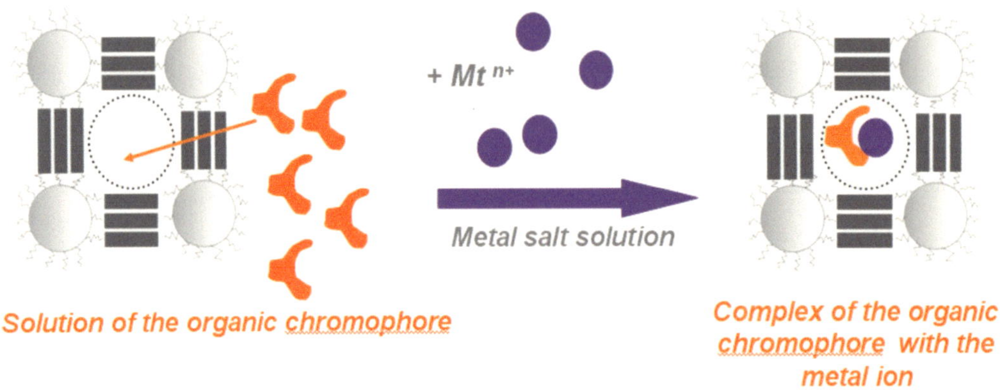

Figure 8. Scheme of the organic chromophore concentration and complex formation with metal ions in mesoporous polymers

In this work, as organic chromophores were used 1-(2-pyridylazo)-2-naphthol (PAN), arsenazo III and phenazo as reactants, soluble salts of copper $CuSO_4$ and cobalt $CoCl_2$, manganese $MnCl_2$, lanthanum $LaCl_3$, calcium $CaCl_2$, magnesium $MgCl_2$ as analytes.

Organic chromophores arsenazo III, phenazo and 1-(2-pyridylazo)-2-naphthol (PAN) have the chemical structure shown in Figure 9.

Immobilization of the organic reagent PAN on mesoporous carriers was carried out by its adsorption from solution in ethanol. It was found that the electron spectrum of PAN on the polymeric carrier was not changed significantly compared with the spectrum of its solution (Figure 10).

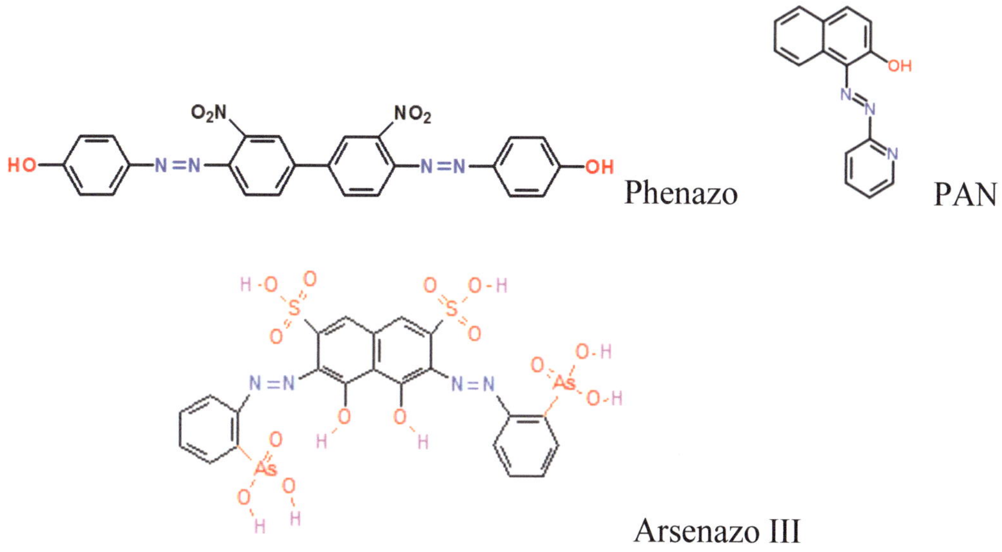

Figure 9. The chemical structure of organic chromophores

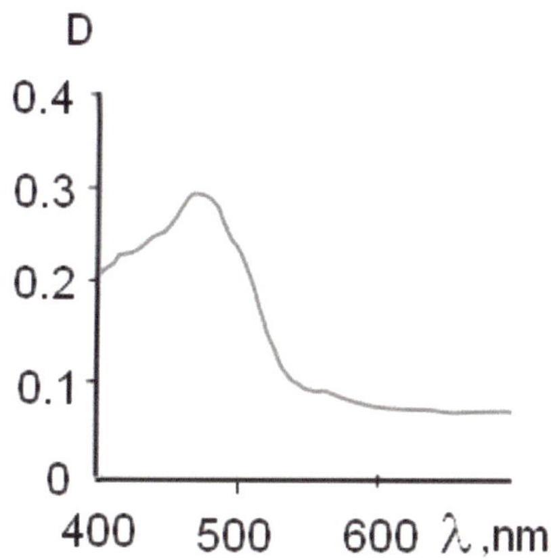

Figure 10. The electronic spectrum of PAN in the mesoporous polymer

To determine the sensitivity limits of complexing reactions of organic reagent with metal cations a series of solutions with salt concentrations 10^{-1} g/l, 10^{-2}, 10^{-3} g/l, 10^{-4} g/l and 10^{-5} g/l was prepared. Polymeric carriers modified with organic reagent were kept in solutions of respective salt for one hour.

Complexes of copper and PAN, PAN and manganese stained polymer in red, complex of cobalt and PAN stained in purple. The value of λ_{max} of PAN complexes with metals when it transferred from solution to mesoporous polymers was not changed. The analysis of the absorption spectra shown in Figures 11 and 13, revealed that 1-(2-pyridylazo)-2-naphthol in mesoporous polymers was able to interact with the metal cations. The height of the characteristic band of the complex PAN - metal depends on the concentration of metal in solution, which further will allow to carry out not only qualitative but also quantitative analysis of metal content.

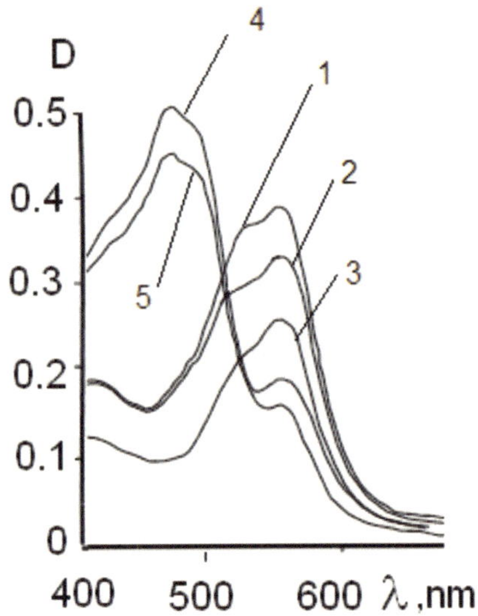

Figure 11. Electronic spectra of the Cu - PAN complex, immobilized in a mesoporous polymer by adsorption from solution,

$[CuSO_4]$= 10^{-1} g/l (1), 10^{-2} g/l (2), 10^{-3} g/l (3), 10-4 g/l (4), 10^{-5} g/l (5)

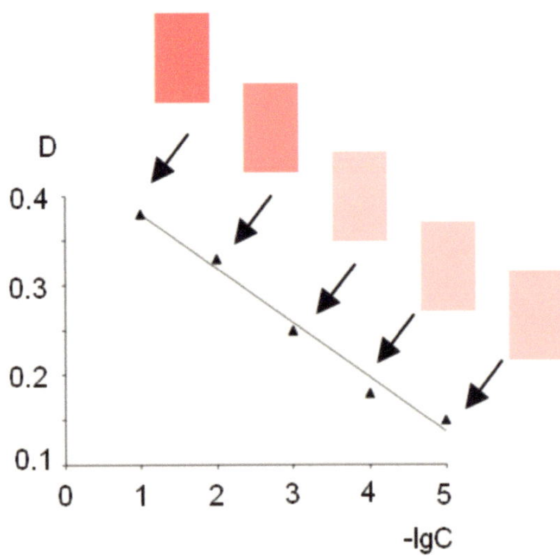

Figure 12. The calibration curve for the Cu - PAN complex in the mesoporous polymer

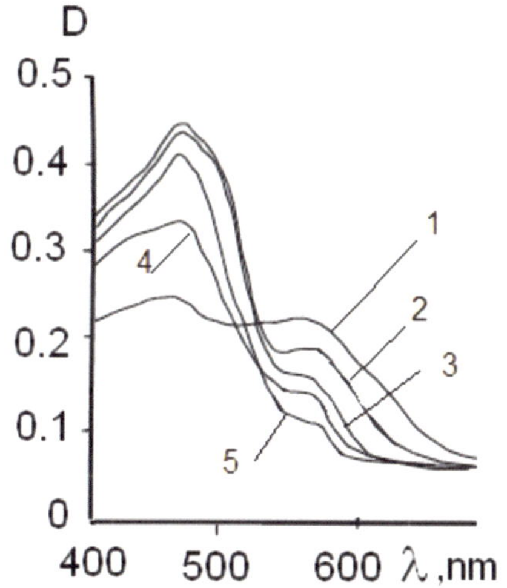

Figure 13. Electronic spectra of the Co - PAN complex, immobilized in a mesoporous polymer by adsorption from solution,

$[CoCl_2]$= 10^{-1} g/l (1), 10^{-2} g/l (2), 10^{-3} g/l (3), 10-4 g/l (4), 10^{-5} g/l (5)

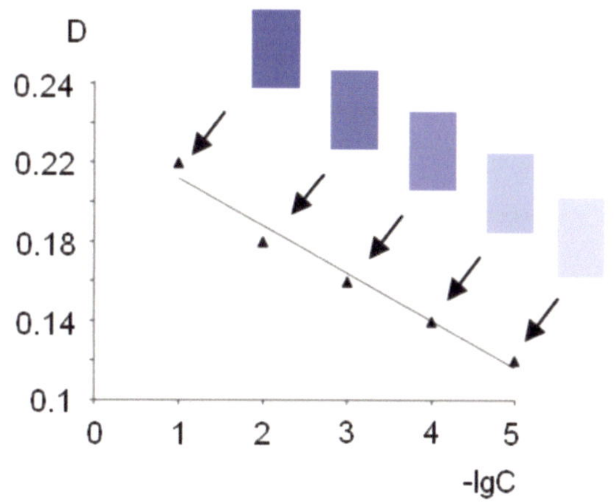

Figure 14. The calibration curve for the Co - PAN complex in the mesoporous polymer

The analysis of the absorption spectra and calibration curves shown in Figures 10 - 14 for 1-(2-pyridylazo)-2-naphthol in mesoporous polymers revealed that the sensitivity of the complexation reaction of PAN and manganese cations on solid carriers was 10^{-5} g/l.

Phenazo chromophore forms with magnesium in alkaline medium an adsorption compound of blue-purple color, the reagent solution is painted in crimson. The absorption maxima of reagent and its complex with magnesium are observed at 490 and 560 nm, respectively. For the reagent and compound with magnesium a molar absorption coefficient is 13900 and 35400, respectively [20]. The optimum concentration of NaOH is 1-2N. Colouring of magnesium compound is stable for 1 hour.

Phenazo's spectra and its complexes with magnesium adsorbed in the pores of the polymer are shown in Figure 15. The limit of sensitivity of complexation reaction of chromophore with magnesium in this case was 10^{-5} g/l.

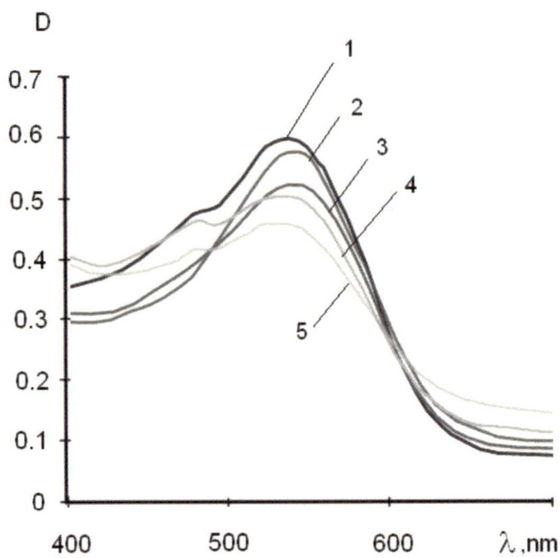

Figure 15. Electronic spectra of the Mg - Phenazo complex, immobilized in a mesoporous polymer by adsorption from solution,

$[MgCl_2]$= 10^{-1} g/l (1), 10^{-2} g/l (2), 10^{-3} g/l (3), 10-4 g/l (4), 10^{-5} g/l (5)

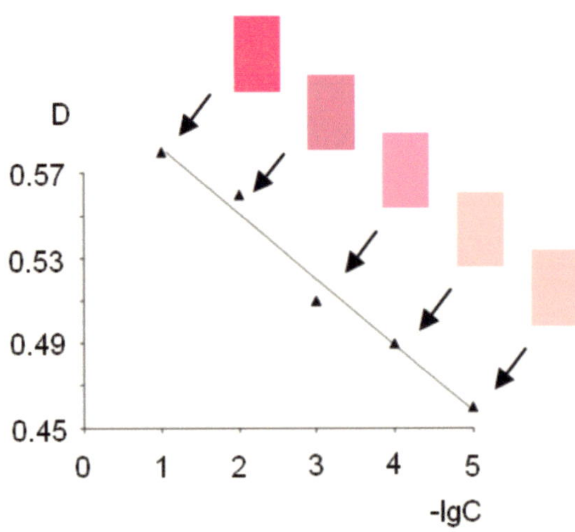

Figure 16. The calibration curve for the Mg - Phenazo complex in the mesoporous polymer

The main feature of the reagent arsenazo III is its ability to form very strong chelates with elements. The good contrast of complexes and large values of molar absorption coefficient ($50{\bullet}10^3$ - $130{\bullet}10^3$) along with the ability to reach high dilutions without dissociation of complexes provides high sensitivity of reactions - up to 0.01 μg/ml.

A maximum of the electronic spectrum of arsenazo III corresponds to a wavelength of 540 nm. From the literature [20], it is known that the absorption spectra of the complex of lanthanum with arsenazo III have the maximum at wavelength of 665 nm and a molar extinction coefficient of the complex 4.5 10^4, the maximum of the absorption spectrum of the complex of calcium and arsenazo III occurs at the wavelength of 655 nm, the molar extinction coefficient of complex is 10^4.

Figure 17 shows the electronic spectrum of mesoporous polymer-modified arsenazo III. The spectrum does not change much compared to the spectrum of an aqueous solution of arsenazo III. Modified films were studied as an analytical sensor for detecting lanthanum and calcium. Figures 18 and 20 show the electronic spectra corresponding to the complexes of arsnazo III with calcium and lanthanum. It was found that complexes of arsenazo III with lanthanum and calcium, immobilized on a polymer, have clearly expressed characteristic bands on the electronic spectra. In case of the complex arsenazo III - calcium the bands at 600 and 655 nm are observed, for the complex arsenazo III - lanthanum the bands appear at 605 and 665 nm.

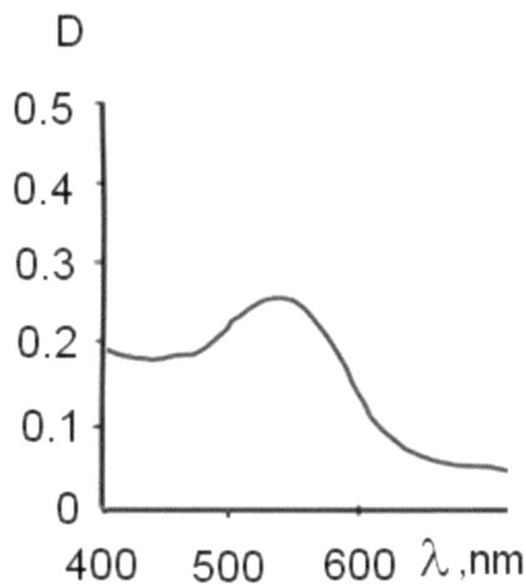

Figure 17. The electronic spectrum of arsenazo III in the mesoporous polymer

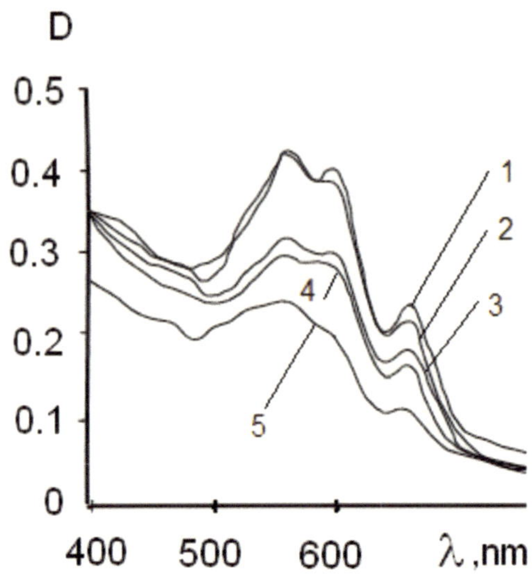

Figure 18. Electronic spectra of the Ca - Arsenazo III complex, immobilized in a mesoporous polymer by adsorption from solution,

$[CaCl_2]$= 10^{-1} g/l (1), 10^{-2} g/l (2), 10^{-3} g/l (3), 10-4 g/l (4), 10^{-5} g/l (5)

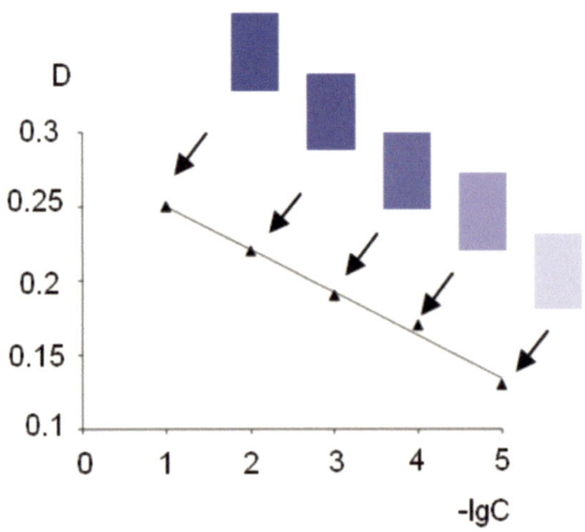

Figure 19. The calibration curve for the Ca – Arsenazo III complex in the mesoporous polymer

http://dx.doi.org/10.5772/57383

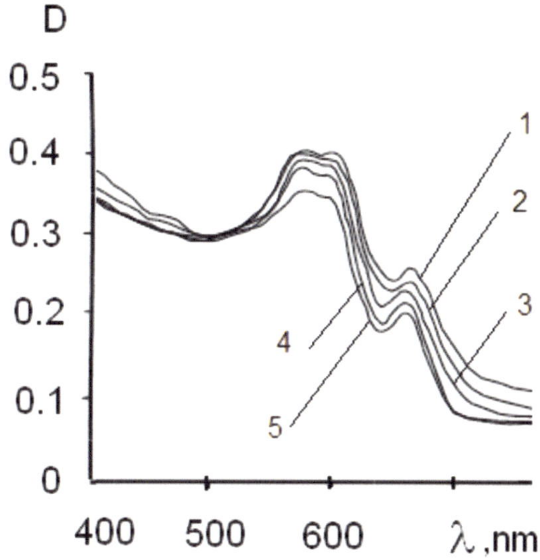

Figure 20. Electronic spectra of the La - Arsenazo III complex, immobilized in a mesoporous polymer by adsorption from solution,

[$LaCl_3$•$6H_2O$]= 10^{-1} g/l (1), 10^{-2} g/l (2), 10^{-3} g/l (3), 10-4 g/l (4), 10^{-5} g/l (5)

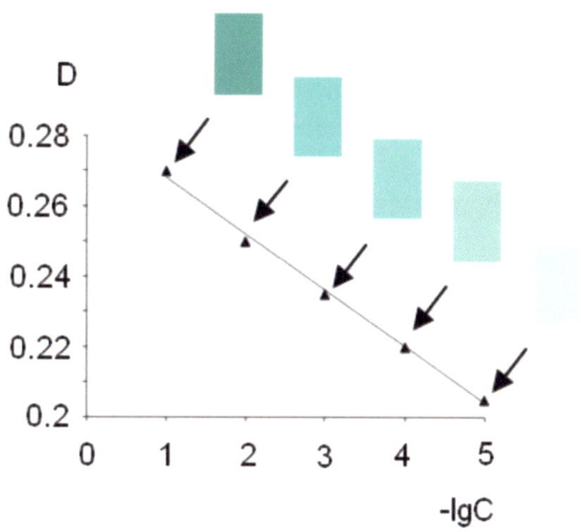

Figure 21. The calibration curve for the La – Arsenazo III complex in the mesoporous polymer

For the polymer matrix modified with arsenazo III, the limit of sensitivity to calcium and lanthanum ions was determined. The maximum concentration at which the metal ions were detected was 10^{-5} g/l.

6. Conclusion

Optically transparent mesoporous block copolymers with regulated free volume were obtained by polyaddition of 2,4-toluene diisocyanate to anionic macroinitiators. It is established that the void formation is conditioned by the geometry of block copolymer self-assembly.

It is established that the chemical nature of a solvent influences the mechanism of polyaddition of 2,4-toluene diisocyanate to the anionic macroinitiator which is a potassium substituted block copolymer of propylene oxide with ethylene oxide.

It is shown that ethyl acetate is the most advantageous solvent for the predominant formation of O-polyisocyanate blocks during the polyaddition of 2,4-toluene diisocyanate to the anionic macroinitiator.

Polymer laser active media based on mesoporous polymers doped by organic luminophores were obtained. The possibility is shown to obtain the induced emission of Rhodamine 6G in mesoporous polymers. It is shown that photochemical stability of organic chemical agent and luminophor rhodamine 6G in polymers made up more than 70,000 pulses.

It was shown that the polymer mesoporous structure provides the possibility to immobilize organic chromatophores in mesoporous polymers. It was established that adsorption processes of organic luminophore of rhodamine 6G were accompanied by the processes of its concentration in a polymer matrix.

The polymer laser-active media based on mesoporous polymers doped by organic chromatophores were obtained. The possibility to obtain the induced emission of radiation of rhodamine 6G in mesoporous polymers was shown. It was also shown that photochemical stability of an organic reagent and luminophore of rhodamine 6G in polymers made up 70.000 pulses.

The qualitative reactions of organic chromatophores arsenazo III, PAN, and phenazo with metal cations on a mesoporous polymer carrier were carried out. It was shown that reaction sensitivity of compex formation of metals and chromatophores on mesoporous polymer substrate made up 10^{-5} g/l.

Acknowledgements

The study was funded by the Russian Foundation for Basic Research, Project No 13-03-97022.

Author details

Ruslan Davletbaev[1*], Alsu Akhmetshina[2], Askhat Gumerov[3] and Ilsiya Davletbaeva[2]

*Address all correspondence to: b210@bk.ru

1 Department of the Materials Science & Technology, Kazan National Research Technical University n.a. A.N. Tupolev, Kazan, Russia

2 Department of the Synthetic Rubber, National Research Technological University, Kazan, Russia

3 Department of the Chemical Cybernetics, National Research Technological University, Kazan, Russia

References

[1] Rouquérol J., Avnir D., Fairbridge C. W., Everett D. H., Haynes J. H., Pericone N., Ramsay J. D. F., Sing K. S. W. Recommendations for the characterization of porous solids. Pure and Applied Chemistry 1994; 8(66) 1739-1758.

[2] Robert W. Cattrall. Chemical Sensors. Oxford University Press; 1997.

[3] Baldini F.; Chester A.N.; Homola J.; Martellucci S. Proceedings of the NATO Advanced Study Institute on Optical Chemical Sensors. Springer; 2006.

[4] Zolotov Y.A., Ivanov V.M., Amelin V.G. Chemical Test Methods of Analysis. Elsevier; 2002.

[5] Amelin V.G., Abramenkova O.I. 2,3,7-Trihydrofluorones on Cellulose Matrices in Test Methods for Determining Rare Elements. Journal of Analytical Chemistry 2008; 11(63) 1112-1120.

[6] Khimchenko S.V., Eksperiandova L.P., Blank A.B. Adsorption-Spectrometric and Test Methods for Determining Perchlorate Ions with Thionine on Polyurethane Foam. Journal of Analytical Chemistry 2009; 1(64) 14-17.

[7] Collinson M.M. Recent Trends in Analytical Applications of Organically Modified Silicate Materials. Trends in Analytical Chemistry 2002; 1(21) 30-38.

[8] Kuznetsov V.V., Sheremet'ev S.V. Analytical Complexation Reactions of Organic Reagents with Metal Ions in a Solidified Gelatin Gel. Journal of Analytical Chemistry 2009; 9(64) 886-895.

[9] Dimosthenis L.G., Evangelos K.P., Mamas I.P., Miltiades I.K. Development of 1-(2-pyridylazo)-2-naphthol-modified Polymeric Membranes for the Effective Batch Pre-

concentration and Determination of Zinc Traces with Flame Atomic Absorption Spectrometry. Talanta 2002; 8(56) 491-498.

[10] Szycher M., editor. Szycher's Handbook of Polyurethanes. CRC Press; 2012.

[11] Puszta Z.I. In situ NMR Spectroscopic Observation of a Catalytic Intermediate in Phosphine – Catalizer Cyclo-Oligomerization of Isocyanates. Angewandte Chemie International Edition 2006; 1(45) 107-110.

[12] Davletbaeva I.M., Shkodich V.F., Ekimova E.O., Gumerov A.M. The Study of Polyisocyanate Groups Opening Initiated by Polyoxyethylene glycolate of Potassium. Journal of Polymer Science Part A: Polymer Chemistry 2007; 8(45) 1494-1501.

[13] Davletbaeva I.M., Shkodich V.F., Gumerov A.M., et al. Intermolecular Interactions in Metal Containing Polymers Based on 2,4-toluene diisocyanate and Open-Chain Analogues of Crown Ethers. Journal of Polymer Science Part A: Polymer Chemistry 2010; 4(48) 592-598.

[14] Davletbaev R.S., Akhmetshina A.I., Avdeeva D.N., Gumerov A.M., Davletbaeva I.M. Investigation of Features of Interaction of Anionic Macroinitiators with 2,4-toluene diisocyanate. Vestnik of Kazan State Technological University Journal 2012; 20 131-133.

[15] Davletbaev R.S., Akhmetshina A.I., Gumerov A.M., Sharifullin R.R., Davletbaeva I.M. The Influence of Solvent Nature on the Mechanism of the Reaction of Anionic Macroinitiators and Aromatic Isocyanates. Butlerov Communications 2013; 9(35) 9-13.

[16] Tager A. Physical Chemistry of Polymers. Moscow: Mir Publishers; 1972.

[17] Savvin S.B., Dedkova V.P., Shvoeva O.P. Sorption-spectroscopic and Test Methods for the Determination of Metal Ions on the Solid-phase of Ion-exchange Materials. Russian Chemical Reviews 2000; 3(69) 187-201.

[18] Costela A., Garcia- Moreno I., Sastre R. Materials for Solid-state Dye Lasers.. In: H. S. Nalwa (ed.) Handbook of Advanced Electronic and Photonic Materials and Devices. San Diego: Academic Press; 2001. p161-208.

[19] Hamdan A.S. Al-Shamiri, Maram T.H. Abou Kana, Azzouz I.M., Badr Y.A. Photophysical properties and quantum yield of some laser dyes in new polymer host. of some laser. Optics & Laser Technology 2009; 4(41) 415-418.

[20] Burger K. Organic Reagents in Metal Analysis. Elsevier Science & Technology; 1973.

Chapter 4

Photonic Crystal Laser Based Gas Sensor

Marcus Wolff, Henry Bruhns, Johannes Koeth,
Wolfgang Zeller and Lars Naehle

Additional information is available at the end of the chapter

http://dx.doi.org/10.5772/57147

1. Introduction

The development of new radiation source technologies has a major impact on the progression of optical trace gas detection [1]. Especially semiconductor diode lasers have proven to be extraordinarily suitable devices for spectroscopic sensors. Their small size and their low acquisition cost are here valuable properties. However, it is particularly advantageous that their emission can spectrally be tuned simply via their operating temperature and operating current. Furthermore, diode lasers can be directly modulated via their injection current. Therefore, they represent particularly suitable radiation sources for photoacoustic spectroscopy because this technique is based on the absorption of modulated radiation and its transformation into a sound wave. As an offset-free technique it enables extremely high detection sensitivity [2].

Continuous-wave (cw) single-frequency diode lasers, like distributed feedback (DFB) lasers, are particularly suitable for spectroscopy because they avoid any cross-sensitivity and enable very selective gas detection [3,4]. DFB devices were originally developed for the telecommunication industry and can conveniently be operated at room temperature. Meanwhile, available emission wavelengths cover the entire near-infrared spectral range (800 nm – 3000 nm). Most recently, DFB lasers operating in the mid-infrared were introduced (> 3000 nm) [5]. The alternative concept of interband cascade lasers (ICL) covers almost the complete mid-infrared from 3 μm to 6 μm [6]. This wavelength range is extraordinarily important for trace gas detection since many molecules have their strong fundamental vibrational absorption bands in this region, enabling extremely high detection limits. These devices close the gap to quantum cascade lasers (QCL), which are currently available with single-frequency emission wavelengths starting around 4 μm (multimode at 3 μm) [7].

However, semiconductor lasers suffer from the considerable weakness that their spectral tuning range is limited to a few nanometers only [4]. As a result of that, the analysis of a gas mixture typically requires the use of a separate laser for each gas component. This makes tunable diode laser spectroscopy expensive and too complex for many applications. A wider spectral tuning range would solve the problem and is, therefore, highly desirable for further practical applications.

This study presents a new photoacoustic gas detection scheme based on a widely tunable coupled-cavity photonic crystal laser operating around 1900 nm. Chapter 2 covers the structure and design of this custom-made room temperature continuous-wave laser followed by measurements of the unique performance characteristics. The new device was designed to enable sensitive detection of water vapor (H_2O). Chapter 3 describes the photonic crystal laser based gas sensor and presents photoacoustic measurements of water spectra. The concluding Chapter 4 summarizes the results.

2. Photonic crystal laser

2.1. Basics

2.1.1. Photonic crystals

Photonic crystals are materials with a spatial periodicity in their dielectric constant. Under certain conditions, photonic crystals can create a photonic bandgap, i.e. a frequency window in which propagation of electromagnetic radiation through the crystal is inhibited. Light propagation in a photonic crystal is similar to the propagation of electrons and holes in a semiconductor. An electron passing through a semiconductor experiences a periodic potential due to the ordered atomic lattice. The interaction between electron and the periodic potential results in the formation of energy bands and bandgaps. It is not possible for the electron to pass through the crystal if its energy falls in the range of the bandgap. However, defects or a break in the periodicity of the lattice can locally abolish the bandgap and give rise to interesting electronic properties [7,8].

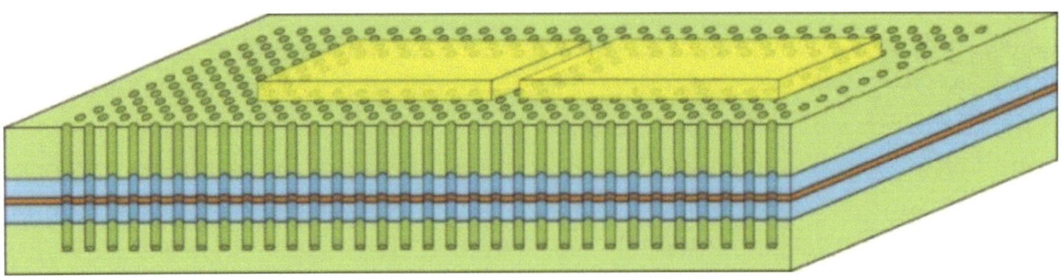

Figure 1. Three-dimensional schematic diagram of photonic crystal structures (courtesy of nanoplus Nanosystems and Technologies GmbH).

Photons in a material with a periodic dielectric constant (index of refraction) experience similar effects. By varying the lattice parameters of these so called photonic crystals, photonic bandgap structures can be obtained that prohibit the propagation of light into the crystal structure. This effect can be used to define optical waveguide structures with very strong lateral light confinement compared to conventional ridge waveguides (of the order of the wavelength) as well as optical filters and wavelength selective reflectors [9,10].

In order for a photon to interact with its periodic environment, its wavelength must be comparable to the periodicity of the lattice. For visible to near-infrared radiation the lattice constant must be in the range of 100 nm to 1 μm. Figure 1 depicts a three-dimensional schematic diagram of photonic crystal structures (courtesy of nanoplus Nanosystems and Technologies GmbH). The yellow blocks on top represent contacts.

2.1.2. Device structure

The photonic crystal laser we are employing in our optical gas sensor is based on an InAs/InGaAs quantum dash-in-a-well structure grown on InP substrate by gas source molecular beam epitaxy. A short period superlattice structure is used for photon confinement. Due to their wide spectral gain bandwidth dash-in-a-well structures are particularly well suited for wavelength tuning applications. The photonic crystal is realized by air-semiconductor structures defined by electron beam lithography and etched into the laser structure by an inductively coupled plasma etch step. The structure comprises a hexagonal lattice with a lattice period of 485 nm and an air fill factor of 30%. The air-filled holes in the semiconductor material exhibit a diameter of 350 nm and a depth of 3 μm. These lattice parameters are chosen so that the middle of the photonic bandgap coincides with the device's emission wavelength. The photonic crystal structures are forming multiple lateral and longitudinal photon confinements as well as two independently contacted laser segments. A waveguide is defined by seven rows of missing holes in a ΓK oriented photonic crystal. The photonic crystal mirror at the rear end of the laser diode is oriented in ΓM direction in order to present a smooth boundary to the traveling light mode, thus, reducing scattering losses. Coupling of the two laser cavities is provided by a photonic crystal coupling section consisting of two rows of etched holes. The output mirror is formed by a single row of holes. A scanning electron micrograph of the rear, middle and front part of the photonic crystal laser fabricated by nanoplus Nanosystems and Technologies GmbH in Gerbrunn, Germany can be seen in Figure 2 [11,12].

The photonic crystal laser integrates two longitudinally coupled Fabry-Pérot (FP) resonators. The main advantage of this principle called coupled cavities is that sophisticated grating structures which are difficult and expensive to implement are avoided. Instead, due to the inverse length dependence of mode spacing in a FP resonator, different mode spacings can be realized by implementing different cavity lengths in each segment. Table 1 lists the lengths and according mode spacings of the two photonic crystal laser segments. The coupled cavity laser diode is emitting radiation in the 1.9 μm wavelength range.

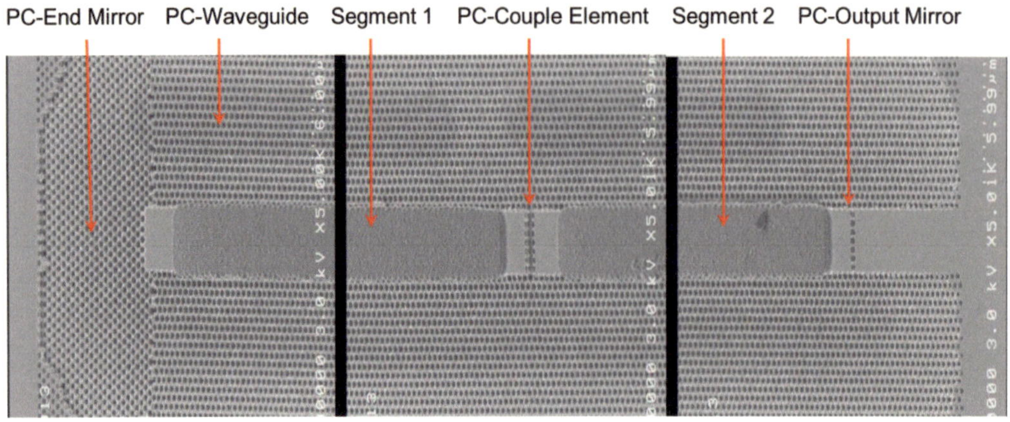

Figure 2. Scanning electron micrograph of the rear, middle and front part of the photonic crystal laser fabricated by nanoplus Nanosystems and Technologies GmbH in Gerbrunn, Germany [11,12].

• segment no.	cavity length	mode spacing
• 1 (long, rear)	336 µm	1.59 nm
• 2 (short front)	241 µm	2.22 nm

Table 1. Cavity lengths and mode spacings of the two photonic crystal laser segments.

2.2. Characterization

The optical output power and the spectral emission of the photonic crystal laser were experimentally investigated. The results are presented in the following subsections. However, not all of the laser parameters important for practical applications have been studied yet. Particularly the temporal stability of wavelength and optical power as well as the sensitivity to environmental factors such as ambient temperature and relative humidity have to be examined.

2.2.1. Emission power

The photonic crystal laser is working in continuous-wave mode (cw). Both laser segments can be operated separately, each with a maximum current of 120 mA. Similarly to traditional diode lasers, the optical output power of the photonic crystal laser depends on temperature and current. The emission power as function of the operating currents of the two laser segments has been measured using the laser diode controller LDC402B/TED 420 from Profile and a Nova II Laser Power Meter with thermal head 3A-FS-SH from Ophir Optronics.

Figure 3 displays the results recorded at an operating temperature of 20°C. It is obvious that the power is a relatively complex function of the two currents, compared to a traditional single cavity laser. The power is not proportional to the currents. Therefore, normalization according to the instantaneous optical emission power is mandatory for the exact measurement of absorption spectra. Otherwise falsifications will occur. The threshold currents of the laser

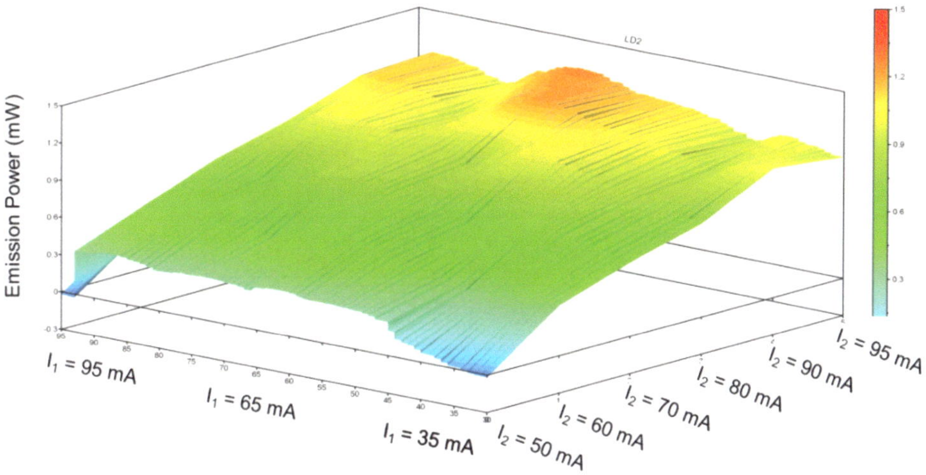

Figure 3. Optical output power of the photonic crystal laser as function of the operating currents of the two laser segments.

segments depend on the operating current of the respective other segment and are between 20 mA and 35 mA for segment 1 and between 35 mA and 50 mA for segment 2, respectively. The laser provides a maximum emission power of approximately 1.3 mW.

2.2.2. Emission wavelength

Similarly to traditional semiconductor lasers, the spectral emission of the photonic crystal device can be tuned by changing the effective index of refraction of the semiconductor material. This can be achieved by varying the temperature or the carrier density. For the purpose of wavelength tuning via temperature, the laser is equipped with a Peltier element (thermoelectric cooler) and a thermo-resistant temperature sensor. Its operating temperature can be adjusted between 10°C and 35°C (i.e. around room temperature). The carrier density is determined by the operating current.

The two independent laser segments are each supporting several discretely spaced lasing modes (see Table 1). Single mode emission occurs if the overlap between two peaks is sufficiently large, i.e. their wavelengths are approximately equal. Independently changing the two segment's effective indexes of refraction results in a shift of the two mode combs relative to each other, which is accompanied by a significant change in the laser emission wavelength. This is known as the Vernier effect [13,14]. The selected segment lengths ensure that a good overlap exists for one pair of laser modes only, for most combinations of electrical currents.

This tuning mechanism for the two coupled cavities of the photonic crystal laser is schematically displayed in Figure 4. The operating current of segment 2, I_2, is successively increased while the operating current of segment 1, I_1, is kept constant. Initially, the "red" mode is emitted because their two counterparts feature the best overlap. Its wavelength can continuously be tuned by current I_2 until the accordance between the "green" modes becomes better than that

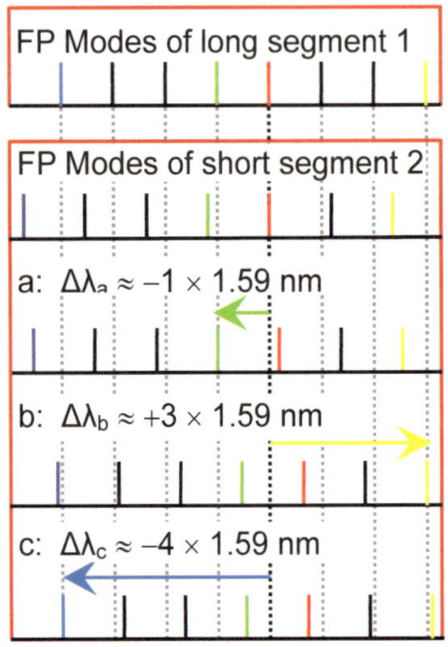

Figure 4. Schematic of the Vernier effect for the two coupled cavities of the photonic crystal laser. I_2 was increased while I_1 was kept constant.

of the "red". Consequently, the mode-hop (a) of $\Delta\lambda_a$ = -1 × 1.59 nm takes place. If current I_2 is further increased, the hop (b) of $\Delta\lambda_b$ = +3 × 1.59 nm to mode "yellow" and, subsequently, hop (c) of $\Delta\lambda_c$ = -4 × 1.59 nm to mode "blue" will occur, each after some continuous tuning. Wavelength tuning with the coupled-cavity photonic crystal laser is, therefore, considered "quasi-continuous" (discretely separated ranges of continuous tuning).

Measurements of the emission wavelength as function of both operating currents were conducted using the laser diode controller LDC402B/TED 420 from Profile and a Fourier transform infrared spectrometer with a DTGS detector (Newport FTIR 80250). Figure 5 shows the emission wavelength as function of I_2 while I_1 is kept constant at 45 mA. Mode-hops (a), (b), and (c) are marked. A similar behavior can be observed if the current of segment 1, I_1, is increased while I_2 is kept constant. Figure 6 shows the results for I_2 = 65 mA. According mode-hops are marked.

Table 2 lists the complete measuring results recorded at an operating temperature of 15°C. Wavelengths are in nm. Empty table cells are affiliated with operating parameters which did not allow single-mode operation. During single-mode operation a sidemode-suppression-ratio (SMSR) larger than 20 dB was achieved.

A total tuning range of more than 20 nm (1900 nm – 1920 nm) was observed which is approximately four times as wide as that of traditional semiconductor lasers. Single-mode emission wavelengths in the spectral proximity of water absorption lines are highlighted in Table 2 (1905.7 nm, 1906.7 nm, 1907.0 nm, 1909.0 nm, 1910.6 nm, 1913.4 nm, 1914.9 nm). The according water vapor absorption spectrum is displayed in Figure 7. It was calculated for a concentration of 1% at room temperature (20°C) and atmospheric pressure (1013 hPa) using the HITRAN database [15]. Absorption lines that are spectrally accessible with the photonic crystal laser are marked. The colors are corresponding to the highlighted cells in Table 2.

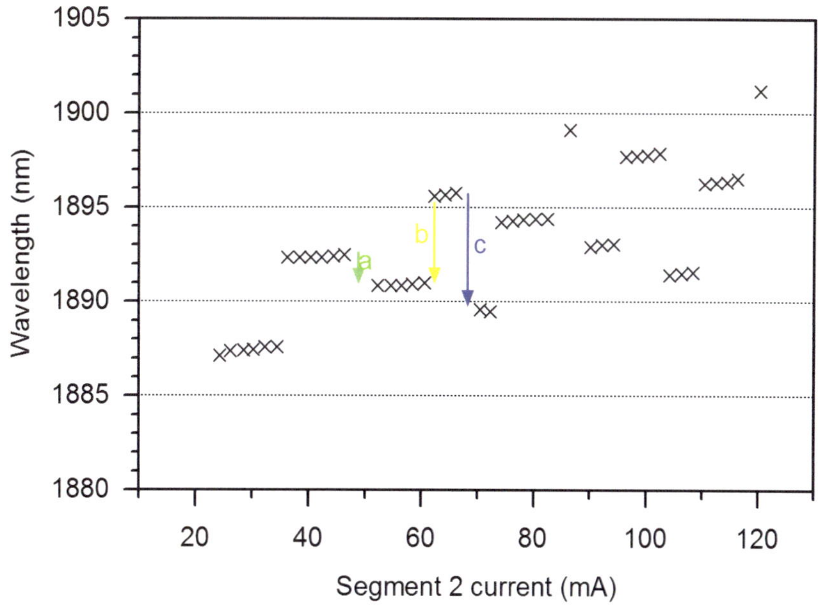

Figure 5. Emission wavelength of the photonic crystal laser as function of segment 2 current, I_2, (I_1 = 45 mA = constant).

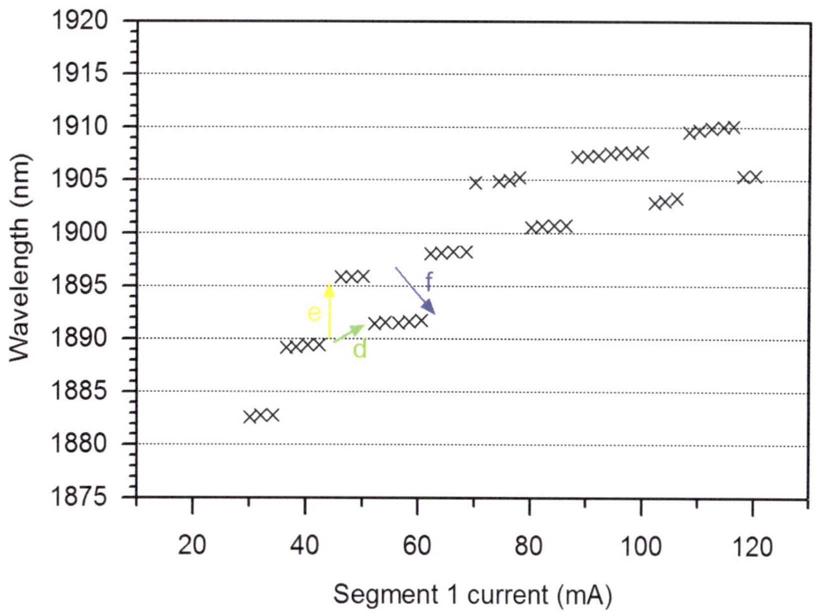

Figure 6. Emission wavelength of the photonic crystal laser as function of segment 1 current, I_1, (I_2 = 65 mA = constant).

Segment 1 current (mA)	35	40	45	50	55	60	65	70	75	80	85	90	95	100
100	1909.00	1909.10	1911.25						1916.05	1916.15		1911.45	1911.60	
95		1910.95	1911.05	1911.15	1906.40	1906.55		1908.80	1915.90		1911.15	1911.35	1913.40	1913.60
90				1906.15	1906.25	1906.40	1908.55			1910.85	1911.00	1913.20	1913.30	
85		1910.60	1905.90	1906.00		1908.25	1908.40		1910.60	1910.80	1912.85	1913.05		1915.30
80		1905.65	1905.75		1907.95	1908.15		1910.35			1912.80		1915.05	1915.25
75	1905.45	1905.55	1905.60	1907.75	1907.85		1910.05		1912.30	1912.50		1914.80	1914.95	
70			1907.50	1907.60		1909.85		1912.05	1912.20		1914.50	1914.65		1916.95
65		1907.30			1909.60			1905.05			1907.45	1909.50	1909.70	
60		1900.20		1909.35	1902.55	1904.65	1904.80		1906.95	1907.20	1909.25	1909.45	1909.60	1911.60
55			1902.25	1902.35	1904.35	1904.55	1906.55	1906.75		1908.95	1909.15		1911.35	1911.55
50		1902.00	1902.05	1904.15	1904.25	1906.30	1906.45	1906.65	1908.70	1908.90	1910.90	1911.05	1911.30	1913.30
45				1904.10	1906.05	1906.25		1908.45	1908.65		1910.85	1910.95		1913.20
40				1905.85	1906.00	1906.15	1908.20	1908.35	1910.35	1910.55	1910.75		1913.00	1913.15
35				1905.80	1907.75	1907.90	1908.10	1910.10	1910.30	1910.50	1912.50	1912.70	1912.95	1913.10
30					1907.70	1907.85	1908.00	1910.05	1910.25	1910.40	1912.45	1912.65	1912.85	1914.90
25					1907.65	1907.85	1909.80	1910.00	1910.20		1912.40	1912.60	1912.85	1914.85
20					1907.60	1907.75	1909.80	1909.90	1910.10	1912.15	1912.35	1912.55	1912.70	1914.80

Segment 2 current (mA)

Empty table cells are affiliated with operating parameters which did not allow single-mode operation. Wavelengths in the spectral proximity of the following water absorption lines are highlighted (nm):

1905.7 1906.7 1907.0 1909.0 1910.6 1913.4 1914.9

Table 2. Emission wavelength in nm as function of the operating currents of the two laser segments recorded at an operating temperature of 15°C [16].

http://dx.doi.org/10.5772/57147

Varying the currents in both segments simultaneously enables fine-tuning of the emission wavelength. A combination of coarse and fine tuning mechanisms enables scanning of larger absorption spectra with a single laser device.

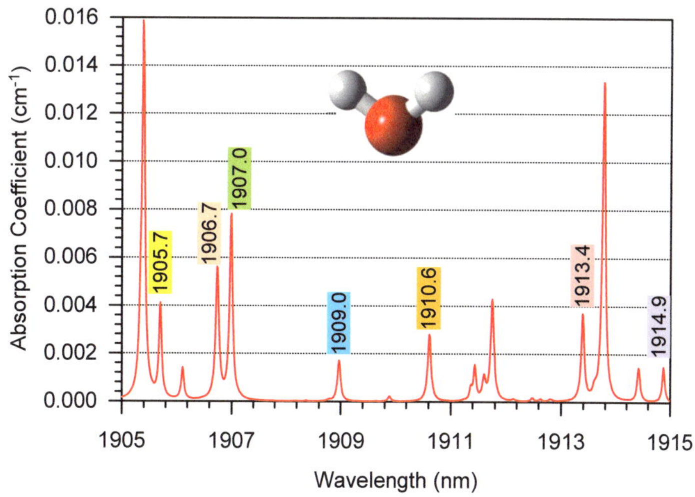

Figure 7. Water vapor absorption spectrum for a concentration of 1% at room temperature (20°C) and atmospheric pressure (1013 hPa) according to the HITRAN database [15]. Absorption lines that are spectrally accessible with the photonic crystal laser are marked.

3. Photoacoustic sensor

3.1. Photoacoustic spectroscopy

The new optical analyzer based on the photonic crystal laser is using photoacoustic spectroscopy. Photoacoustic or optoacoustic spectroscopy (PAS / OAS) is based on the photoacoustic effect that was discovered in 1880 by Alexander Graham Bell [17]. One year later, Wilhelm Conrad Roentgen published a paper on the application of photoacoustic spectroscopy on gas [18].

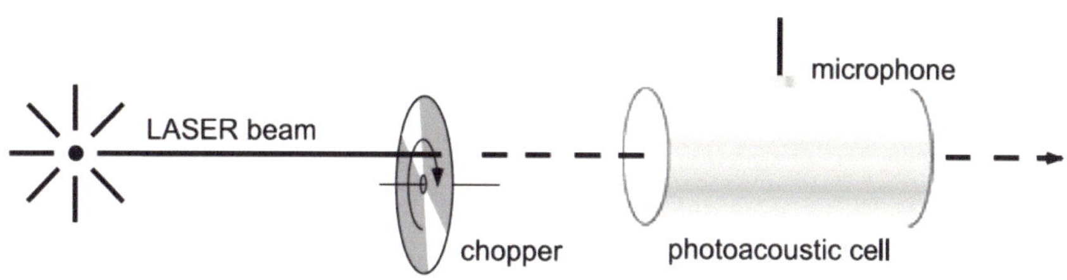

Figure 8. Schematic setup for photoacoustic spectroscopy.

Just like all other spectroscopic techniques photoacoustics is based on the interaction between matter and electromagnetic radiation [1]. If a sample is irradiated with resonant electromagnetic radiation, i.e. the light frequency coincides with vibrational or rotational eigenmodes of the irradiated molecules, photons are absorbed and a fraction of the molecules is excited into an energetically elevated state.

After the so-called lifetime of the excited state the molecules relax, radiantly or non-radiantly, back into their ground state. PAS is using the part of the energy that is transferred into kinetic energy of collision partners via inelastic collisions (non-radiant relaxation). The velocity of the colliding gas molecules increases. However, an increase of the molecule's average velocity is equivalent with an increase of the temperature of the gas. As long as the sample can be approximately considered an ideal gas, a higher temperature results in a thermo-elastic expansion of the sample which is, at constant volume, equivalent to a local increase of the pressure.

If the irradiation of the sample is interrupted, heat conduction and shear stress lead to a quick deduction of the thermal energy via cell walls and gas volume and, therefore, to a reduction of the pressure back to the initial value. Modulated radiation results in a small periodic pressure variation (typically 10^{-6} to 10^{-3} Pa) with the modulation frequency of the radiation source, the so-called photoacoustic signal. The detection of this sound wave by a microphone eliminates the offset (usually the atmospheric pressure), since only changes of pressure are detected. The usually following phase-sensitive measurement of the microphone signal with a lock-in amplifier generates a constant signal [2].

The schematic experimental set-up for photoacoustic spectroscopy is shown in Figure 8. It displays a laser whose radiation is mechanically modulated with a chopper and guided through the sample cell that contains the microphone.

Since a photoacoustic signal is produced only when light is absorbed, PAS is, contrary to transmission spectroscopy, considered an offset-free technique. The photoacoustic signal, S, is directly proportional to the concentration of absorbing molecules, ϱ, the absorption cross-section of the molecular transition, σ, and the optical output power of the radiation source, P_0,

$$S = \varrho \cdot \sigma \cdot P_0 \cdot C_{CELL} \tag{1}$$

The proportionality factor, C_{CELL}, is often called the cell constant because it primarily depends on the geometry of the sample cell. To a certain extent, however, it is also a function of the beam profile and position, the microphone sensitivity and the modulation function of the light source. The cell constant and, therefore, the sensitivity of a photoacoustic sensor can be considerably improved by taking advantage of acoustical resonances of the sample cell. Modulating the laser radiation at a frequency equivalent to an acoustical mode of the cell enables an enormous enhancement of the signal [19].

Gas sensors based on the photoacoustic effect allow the detection of extremely low concentrations. Detection limits of the order of parts-per-trillion are achievable [2]. Using spectrally fine radiation photoacoustic gas detection is also a very selective method. It is even possible

http://dx.doi.org/10.5772/57147

to discriminate different isotopes of one molecule [20]. With a spectrally tunable single-frequency radiation source the absorption coefficient α of the absorbing gas component can directly be measured as function of the wavelength $\alpha(\lambda)$ [21]. We applied this method and recorded water vapor spectra within the continuously tunable wavelength ranges of the photonic crystal laser.

In summary, photoacoustics is an extremely powerful technique and its field of application reaches from gas detection for medical diagnostics [22] to imaging with photoacoustic tomography [23] and is still growing.

3.2. Experimental setup

Figure 9 depicts the experimental setup of the photonic crystal laser based gas sensor. Temperature and electrical currents for the two laser segments are controlled using the laser diode controller LDC402B/TED 420 from Profile. The radiation is mechanically modulated by a high-precision chopper (Scitec Instruments; type 300 CD). After passing the sample cell, the optical power is measured by a thermopile detector (Ophir Laser Measurement Group; type NOVA II with thermal head 3A-FS-SH). Due to the very small absorptions of water vapor in the investigated spectral region the signal qualifies for the power normalization of the photoacoustic signal.

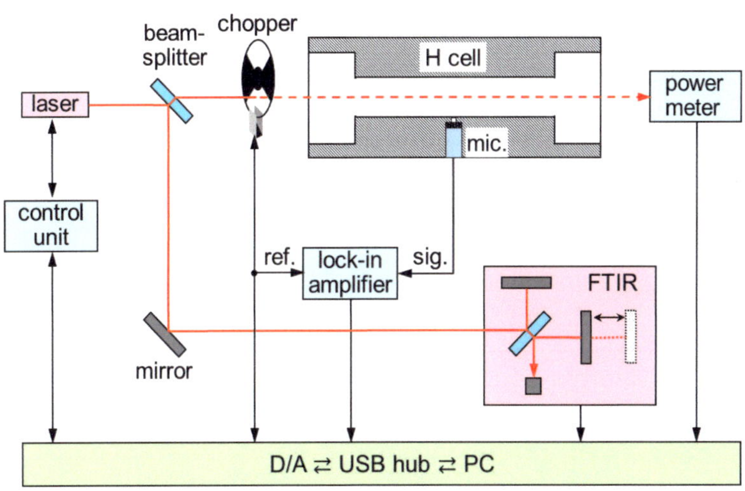

Figure 9. Experimental setup of the photonic crystal laser based gas sensor.

The measuring chamber features the well-established H geometry [24]. It combines a thin cylindrical resonator of diameter $d_1 = 6$ mm and length $l_1 = 60$ mm with two cylindrical buffer volumes of diameter $d_2 = 24$ mm and length $l_2 = 30$ mm at the ends. The overall absorption length of the cell is 120 mm. Its first longitudinal mode has a resonance frequency of approximately 2,800 Hz which was chosen as the modulation frequency. The cell is made of Teflon (PTFE) and the windows are of CaF_2.

The photoacoustic signal is detected by an electret microphone (Primo Microphones GmbH; EN158N), pre-amplified (PAS-Analytik; Amp1) and phase-sensitively recorded (Signal Recovery; Model 7265). The lock-in amplifier uses the output signal of the chopper as reference signal. It measures the absolute value of the photoacoustic signal with a time constant of 1 s. A CaF_2 beamsplitter separates a small fraction of the radiation for spectral characterization with a Fourier transform infrared spectrometer (FTIR; MIR™ 8025 Model 80250 manufactured by Newport Corporation).

3.3. Measurements

Using the experimental setup described in the previous subsection, photoacoustic spectra of ambient air at atmospheric conditions (1% H_2O, 20°C, 1013 hPa) were recorded. At four constant operating currents of laser segment 2 (I_2 = 50 mA / 55 mA / 60 mA / 75 mA), the current of laser segment 1 was tuned as far as continuously (mode-hop free) possible. The laser temperature was held constant at 20°C. Converting current values into corresponding emission wavelengths using previously acquired data (see e.g. Table 2) wavelength dependent absorption plots were generated. The variation in laser output power was compensated by means of normalization according to the instantaneous optical emission power.

Figure 10 displays the four spectra, each one of a single absorption line. As can be seen, the measured photoacoustic signal (crosses) closely tracks the absorption data for water vapor as provided by the HITRAN database [15] (solid line). No cross-sensitivities with other components of air occurred.

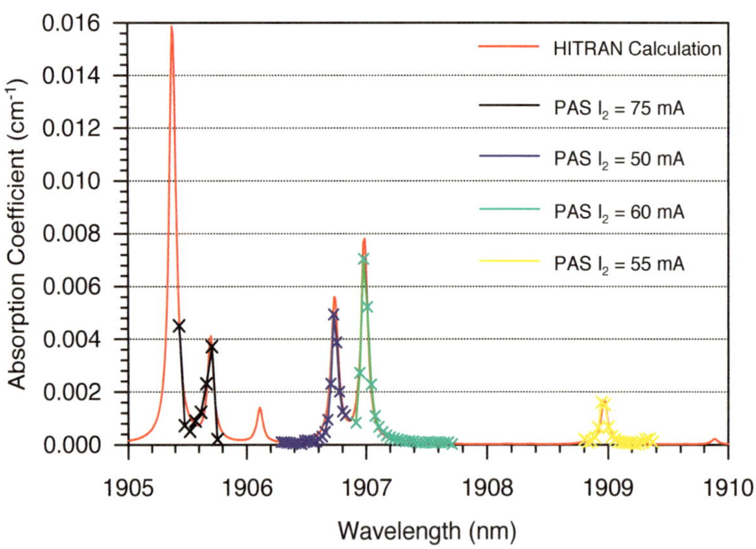

Figure 10. Four photoacoustic spectra of ambient air at atmospheric conditions (1% H_2O, 20°C, 1013 hPa), recorded at constant operating current of laser segment 2 (I_2) by tuning the operating current of segment 1 (I_1) together with a water vapor absorption spectrum according to the HITRAN database [15].

4. Conclusion

We presented a new optical gas detection scheme based on a photonic crystal laser. We showed emission and operation parameters of this radiation source as well as measured spectra of H_2O at 1910 nm.

The optical gas sensor based on the photonic crystal laser provides some considerable advantages over those based on traditional semiconductor lasers. The wide spectral tunability of more than 20 nm allows multigas detection using only one laser. Since the emission wavelength is primarily determined by the two laser currents, its spectral tuning is very fast. Therefore, the analyzer provides a very short response time. Furthermore, the photonic crystal high-reflectance end mirror enables high emission power, which results in combination with photoacoustic spectroscopy in high detection sensitivity. Since these devices do not require the integration of a wavelength selective grating for single-mode emission, it is expected that they will become considerably cheaper than traditional DFB lasers [25,26]. Next generation photonic crystal lasers will feature an even larger tuning range and a higher optical output power of approximately 20 mW. This will qualify these devices even better for a quick and highly sensitive multigas detection using a single laser source.

Author details

Marcus Wolff[1*], Henry Bruhns[1], Johannes Koeth[2], Wolfgang Zeller[2] and Lars Naehle[2]

*Address all correspondence to: marcus.wolff@haw-hamburg.de

1 Hamburg University of Applied Sciences, School of Engineering and Computer Science, Department of Mechanical Engineering and Production, Heinrich Blasius Institute for Physical Technologies, Hamburg, Germany

2 Nanoplus Nanosystems and Technologies GmbH, Gerbrunn, Germany

References

[1] Demtroeder W. Laser spectroscopy. 4th ed. Berlin: Springer; 2009.

[2] Michaelian KH. Photoacoustic Infrared Spectroscopy. 2nd ed. Hoboken (NJ): Wiley-Interscience; 2011.

[3] Fehér M, Jiang Y, Maier JP. Optoacoustic trace-gas monitoring with near-infrared diode lasers. Applied Optics 1994;33(9):1655-1658.

[4] Wolff M, Harde H. Photoacoustic Spectrometer based on a DFB-Diode Laser. Infrared Physics and Technology 2000;41(5):283-286.

[5] Naehle L, Belahsene S, von Edlinger M, Fischer M, Boissier G, Grech P, Narcy G, Vicet A, Rouillard Y, Koeth J, Worschech L. Continuous-wave operation of type-I quantum well DFB laser diodes emitting in 3.4 μm wavelength range around room temperature. Electronics Letters 2011;47(1):46-47.

[6] Naehle L, Hildebrandt L, Kamp M, Hoefling S. Interband Cascade Lasers: ICLs open opportunities for mid-IR sensing. Laser Focus Word 2013;49(5):70-73.

[7] 2012 Buyer's Guide. Laser Focus Word 2012;48(3):45-46, 61, 64.

[8] Joannopoulos JD. Photonic Crystals: Molding the Flow of Light. 2nd ed. Princeton (NJ): University Press; 2008.

[9] Skorobogatiy M, Yang J. Fundamentals of Photonic Crystal Guiding. Cambridge (MA): University Press; 2008.

[10] Mahnkopf S, Maerz R, Kamp M, Duan GH, Lelarge F, Forchel A. Tunable photonic crystal coupled-cavity laser. IEEE Journal of Quantum Electronics 2004;40:1306-1314.

[11] Mueller M, Scherer H, Lehnhardt T, Roessner K, Huemmer M, Werner R, Forchel A. Widely tunable coupled cavity lasers at 1.9 μm on GaSb. IEEE Photonics Technology Letters 2007;19:592-594.

[12] Naehle L, Zimmermann C, Zeller W, Bruckner K, Sieber H, Koeth J, Hein S, Hofling S, Forchel A. Widely tunable photonic crystal coupled cavity laser diodes based on quantum-dash active material. Proceedings of the 20th International Conference on Indium Phosphide & Related Materials, IPRM 09, 10-14 May 2009; Newport Beach, CA. Red Hook (NY): Curran Associates; 2009. pp.28-30. http://dx.doi.org/10.1109/ICIPRM.2009.5012418

[13] Zeller W, Legge M, Seufert J, Werner R, Fischer M, Koeth J. Widely tunable laterally coupled distributed feedback laser diodes for multispecies gas analysis based on InAs/InGaAs quantum-dash material. Applied Optics 2009;48:B51-B56.

[14] Bauer A, Mueller M, Lehnhardt T, Roessner K, Huemmer M, Hofmann H, Kamp M, Hoefling S, Forchel A. Discretely tunable single-mode lasers on GaSb using two-dimensional photonic crystal intracavity mirrors. Nanotechnology 2008;19:235202.

[15] Heinrich J, Langhans R, Seufert J, Hofling S, Forchel A. Quantum cascade microlasers with two-dimensional photonic crystal reflectors. IEEE Photonics Technology Letters 2007;19(23):1937-1939.

[16] Rothman LS, et al. The HITRAN2012 molecular spectroscopic database. Journal of Quantitative Spectroscopy and Radiative Transfer 2013 (in press). http://dx.doi.org/10.1016/j.jqsrt.2013.07.002i

[17] Datasheet 1900nm PK-Laser 362-6-18. nanoplus Nanosystems and Technologies GmbH; 2008.

[18] Bell AG. On the Production and Reproduction of Sound by Light. American Journal of Science 1880;20:305-309.

[19] Roentgen WC. Versuche ueber die Absorption von Strahlen durch Gase; nach einer neuen Methode ausgeführt. Bericht der Oberhessischen Gesellschaft für Natur- und Heilkunde 1881;XX:52-58.

[20] Wolff M, Baumann B, Kost B. Shape-optimized photoacoustic cell: Experimental confirmation and numerical consolidation. Int. J. Thermophysics 2012;33(10):1953-1959.

[21] Wolff M, Harde H, Groninga H. Isotope-Selective Sensor based on PAS for Medical Diagnostics. Journal de Physique IV 2005;125:773-775.

[22] Wolff M, Rhein S, Bruhns H, Nähle L, Fischer M, Koeth J. Photoacoustic Methane Detection using a novel DFB-type Diode Laser at 3.3 μm. Sensors and Actuators B: Chemical 2013;187:574-577.

[23] Germer M, Wolff M, Harde H. Photoacoustic NO Detection for Asthma Diagnostics. Proc. SPIE 2009;7371:73710Q.

[24] Wang LV, Hu S. Optoacoustics/Deep Tissue Imaging: Photoacoustic tomography is ready to revolutionize. Science 2012;335:1458–1462.

[25] Nodov E. Optimization of resonant cell design for optoacoustic gas spectroscopy (H-type). Applied Optics 1978;17:1110-1119.

[26] Wolff M, Gebhardt S. Photonic crystal laser for photoacoustic spectroscopy. Proceedings of the 15th International Conference on Photoacoustic and Photothermal Phenomena, ICPPP15, 19-23 Jul 2009, Leuven, Belgium. Red Hook (NY): Curran Associates; 2009. p.323.

[27] Wolff M, Koeth J, Zeller W, Gebhardt S. Photonic crystal laser based gas detection. Proceedings of the 4th EPS-QEOD Europhoton Conference, 29 Aug – 03 Sep 2010, Hamburg, Germany. Mulhouse, France: European Physical Soc; 2010. p.30.

Chapter 5

Microsphere and Fiber Optics based Optical Sensors

A. Rostami, H. Ahmadi, H. Heidarzadeh and
A. Taghipour

Additional information is available at the end of the chapter

http://dx.doi.org/10.5772/57465

1. Introduction

Great advantages of optical sensing including immunity to electromagnetic interference, large bandwidth, reliability, and high sensitivity have interested engineers to apply optical sensors instead of electronic sensors for sensing of various environmental parameters. Optical sensors work based on change of intensity, phase, polarization, wavelength and spectral distribution of the light beams by the phenomenon that is being measured. Optical sensors can be used in many areas such as biomedical, civil and aerospace engineering, oil and gas industry, transportation as well as in energy sector. This chapter provides enough information related to optical sensors that is utilized in biomedical engineering especially. First, WGM-based optical biosensors are introduced and studied. Then microsphere resonator is considered as an example and a specific type of WGM-based biosensors. Helmholtz equation is solved for this micro optical-resonator in order to obtain optical modes and Eigen values. Then, capability of microsphere resonator as biosensor is investigated and analyzed based on simulation and comparison to experimental results. Furthermore, fiber optic sensors and their application in biomedical field, specifically in minimally invasive surgeries, are described.

2. Optical biosensors based on WGM (Whispering Gallery Mode)

There are different types of biosensors with diverse physical basics such as electrochemical biosensors [1] , mass-based biosensors [2] , optical biosensors [3] , etc., among these different types of biosensors, optical biosensors provide more efficiency because of their exclusive features due to light-based detection. One kind of optical biosensor is whispering gallery mode (WGM)-based biosensor which provides label-free detection [4]. WGM phenomenon occurs in optical microcavities (optical micro resonators) [5]. Optical microcavities confine light in a circular path and make resonance phenomenon in specific wavelengths [6, 7]. Therefore, the resonance wavelengths emerge as notches in transmission spectrum. There are various types of optical microcavities according to their shapes. Three major types of optical microcavities can be seen in Fig. 1.

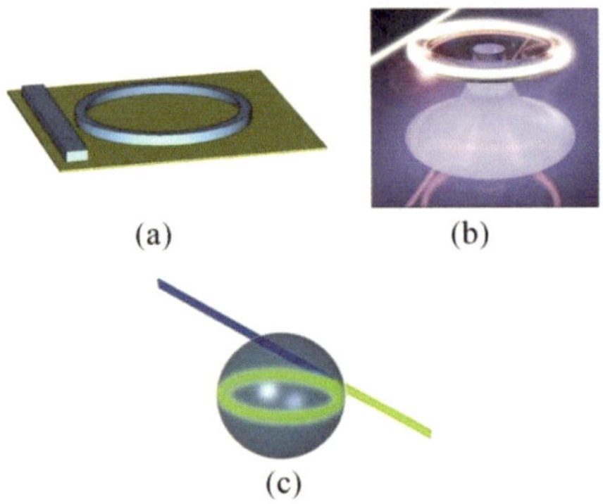

Figure 1. Different types of optical microcavities, (a) Micro-ring resonator, (b) Micro-toroid resonator, (c) Micro-sphere resonator.

Despite easy fabrication process, the ring resonator has small quality factor (Q) compared to micro toroid and microsphere [5, 6]. Specifically, microspheres are three-dimensional WGM resonators, a few hundred micrometers in diameter, often fabricated by simply melting the tip of an optical fiber. The total optical loss experienced by a WGM in this type of optical micro cavity can be extremely low (quality factor (Q) as high as 10^8 are routinely demonstrated) [8]. These extraordinary Q-values translate directly to high energy density, narrow bandwidth for resonant-wavelength and a lengthy path of cavity ring. Because of these advantages, micro-sphere resonators are prone structures for being competent optical biosensors. Fig. 2 shows a microsphere that is coupled to an optical fiber and occurrence of WGM phenomenon.

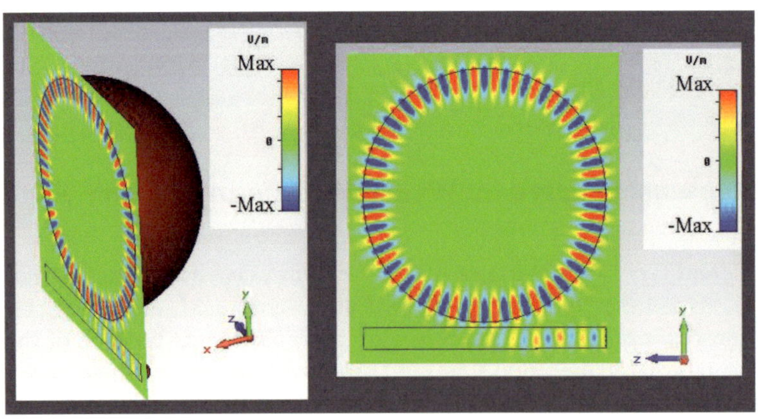

Figure 2. WGM phenomenon in microsphere

For analytical description of WGM in microsphere, the Helmholtz equation should be solved in spherical coordinate.

2.1. Solution of Helmholtz equation for microsphere optical resonator

In order to theoretically investigate the performance of microsphere, the Maxwell wave equation in microsphere should be solved. For solving the wave equation, the separation of variables technique can be used in order to separate the time variable from spatial variables,

$$u(\vec{r},t)=\psi(\vec{r})T(t) \tag{1}$$

The part of the solution related to spatial coordinates, $\psi(\vec{r})$, satisfies Helmholtz equation as follows.

$$\nabla^2\psi+k^2\psi=0 \tag{2}$$

where k^2 is a separation constant. The Helmholtz equation in spherical coordinates is:

$$\frac{1}{r^2}\frac{\partial}{\partial r}\left(r^2\frac{\partial\psi}{\partial r}\right)+\frac{1}{r^2\sin\theta}\frac{\partial}{\partial\theta}\left(\sin\theta\frac{\partial\psi}{\partial\theta}\right)+\frac{1}{r^2\sin^2\theta}\frac{\partial^2\psi}{\partial\phi^2}+k^2\psi=0 \tag{3}$$

If we assume that the polarization is constant throughout all space, the separation of variables method can be utilized for solving Helmholtz equation. For applying this method, we consider the solution given by

$$\psi(r,\theta,\phi)=R(r)\Theta(\theta)\Phi(\phi) \tag{4}$$

where $R(r)$, $\Theta(\theta)$ and $\Phi(\varphi)$ are the radial, polar and azimuthal parts of solution respectively. By applying the considered solution in Helmholtz equation and multiplying by $\frac{r^2}{R\Theta\Phi}$, the following equation is obtained:

$$\frac{1}{R}\frac{d}{dr}\left(r^2\frac{dR}{dr}\right)+k^2r^2+\frac{1}{\Theta\sin\theta}\frac{d}{d\theta}\left(\sin\theta\frac{d\Theta}{d\theta}\right)+\frac{1}{\Phi\sin^2\theta}\frac{d^2\Phi}{d\phi^2}=0 \tag{5}$$

Multiplying Eq. (5) by $\sin^2\theta$, the last term should satisfy

$$\frac{1}{\Phi}\frac{d^2\Phi}{d\phi^2}=-m^2 \tag{6}$$

The solution is

$$\Phi(\phi) = e^{\pm im\phi} \tag{7}$$

where m is integer because of periodic boundary condition. Inserting Eq. (7) into Eq. (5) the following equation can be obtained.

$$\frac{1}{R}\frac{d}{dr}\left(r^2\frac{dR}{dr}\right)+k^2r^2+\frac{1}{\Theta\sin\theta}\frac{d}{d\theta}\left(\sin\theta\frac{d\Theta}{d\theta}\right)-\frac{m^2}{\sin^2\theta}=0 \tag{8}$$

The third and fourth terms in Eq. (8) only depend on θ, while the first and second terms only depend on r. Thus the polar dependence satisfies:

$$\frac{1}{\Theta\sin\theta}\frac{d}{d\theta}\left(\sin\theta\frac{d\Theta}{d\theta}\right)-\frac{m^2}{\sin^2\theta}=-l(l+1) \tag{9}$$

where the separation constant is considered as $l(l+1)$ and l=0, 1, 2, ..., so that the solution is to be finite at θ=0, π. With inserting x=cosθ, the equation (9) changes to equation (10) which is Associated Legendre equation.

$$\frac{d}{dx}\left[\left(1-x^2\right)\frac{d\Theta}{dx}\right]+\left(l(l+1)-\frac{m^2}{1-x^2}\right)\Theta(x)=0 \tag{10}$$

The solution of Eq. (10) is

$$\Theta(x)=P_l^m(x) \qquad (x=\cos\theta) \tag{11}$$

where $P_l^m(\cos\theta)$ are Associated Legendre polynomials and $|m| \le l$.

Usually the polar and azimuthal parts of the solution in Eq. (4) are combined into a spherical harmonic, $Y_l^m(\theta, \phi)$, where

$$Y_l^m(\theta,\phi)=CP_l^m(\cos\theta)e^{im\phi} \tag{12}$$

where C is normalization constant.

From Eq. (8) and (9), the radial dependence in Eq. (4) is given by

http://dx.doi.org/10.5772/57465

$$r^2 \frac{d^2R}{dr^2} + 2r\frac{dR}{dr} + \left[k^2r^2 - l(l+1)\right]R = 0 \tag{13}$$

Inserting $Z(r) = R(r)(kr)^{1/2}$ into Eq. (13), we obtain

$$r^2 \frac{d^2Z}{dr^2} + r\frac{dZ}{dr} + \left[k^2r^2 - (l+1/2)^2\right]Z = 0 \tag{14}$$

which is the Bessel equation of order $l+1/2$. The solutions of Eq. (14) are $J_{l+1/2}(kr)$ and $N_{l+1/2}(kr)$. The solutions for $R(r)$ are the spherical Bessel and Neumann functions, $j_l(kr)$ and $n_l(kr)$, that can be expressed by $J_{l+1/2}(kr)$ and $N_{l+1/2}(kr)$ as below:

$$\begin{aligned} j_l(x) &= \sqrt{\frac{\pi}{2x}} J_{l+1/2}(x) \qquad (a) \\ n_l(x) &= \sqrt{\frac{\pi}{2x}} N_{l+1/2}(x) \qquad (b) \end{aligned} \tag{15}$$

Although the mathematical solution for $R(r)$ consists of Neumann functions, the physical solution does not because the answer is finite in the region of solution. Hence, the general solution of Helmholtz equation in spherical coordinates is

$$\psi(r,\theta,\phi) = \sum_k \sum_{l=0}^{\infty} \sum_{m=-l}^{l} a_{klm} j_l(kr) Y_l^m(\theta,\phi) \tag{16}$$

where a_{klm} are determined by boundary conditions.

Solving Helmholtz equation for light in spherical coordinate and using boundary condition across the surface of microsphere, the characteristic equation is obtained as Eq. (17) that describes the relationship between wave vector and eigenvalues l and m [9].

$$(\eta_s \alpha_s + 1/R_s) \times j_l(kn_s R_s) = kn_s j_{l+1}(kn_s R_s) \tag{17}$$

where η_s is 1 for TE mode and n_s^2 / n_0^2 for TM mode. n_s and n_0 are refractive indices of sphere and surrounding medium, respectively. Also, R_s is the radius of microsphere. In Eq. (17), $\alpha_s = (\beta_l^2 - k^2 n_0^2)^{1/2}$ where $\beta_l = (l^2 + l)^{1/2} / R_s$.

2.2. Bio-sensor based on microsphere optical resonator

Virus particles are a major cause for human diseases, and their early detection is of great urgency since modern day travel has enabled these diseases agents to be spread through population across the globe [10]. Recently, single-virus detection has been reported based on microsphere resonator, practically [11]. Microsphere resonators are used for biosensing applications on the basis of resonant frequency shift due to attachment of desirable particles to the microsphere surface. There are many cases which microspheres are used as a high sensitive biosensor for detection of particles based on resonance frequency shift [12, 13].

An optical WGM may be represented by a light wave that circumnavigates near the surface of a glass sphere. Attachment of a particle to the surface of microsphere causes shift in resonance frequency of transmission spectrum due to perturbation in path of light [14, 15]. Consequence of this variation in the travelling path of light is satisfaction of condition related to WGM in the other frequency which advents as shift of resonance frequency. It is obvious from Fig. 2 that in resonance frequency, the light passing in the fiber will be trapped in microsphere, so this frequency will be omitted from transmission spectrum of optical fiber. It is noteworthy to mention that this figure has been obtained from simulation performed by Finite Element Method (FEM). Transmission spectrum of microsphere resonator without particle attachment is shown in Fig. 3. The obtained Q for this microsphere resonator is calculated as 8.8×10^4. This amount of Q is far small from ultimate Q for microsphere which is almost 10^9 [16]. This smallness of Q for simulated microsphere is because of small amount of considered diameter. In order to increase the quality factor, size of the microsphere have to be increased, but it decreases capability of single small-size particle detection. So, we have to compromise between these two parameters in practice.

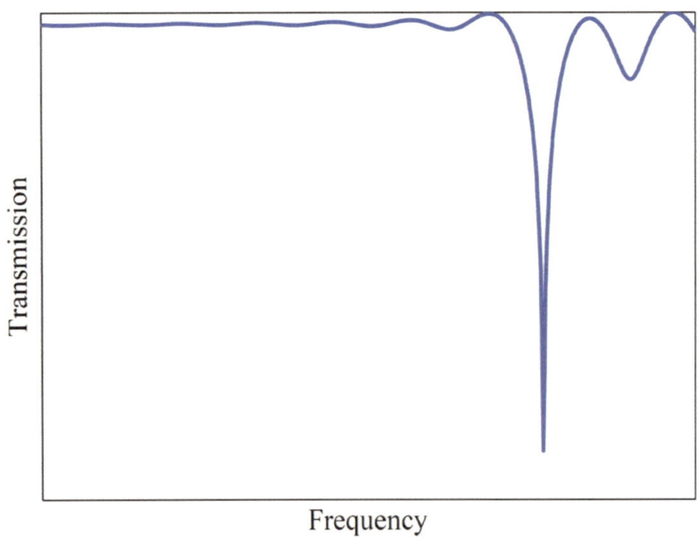

Figure 3. Transmission spectrum of microsphere biosensor without particle attachment [17].

Fig. 4 shows the result of attachment a particle to the surface of the microsphere. It causes a shift of resonance notch in the transmission spectrum.

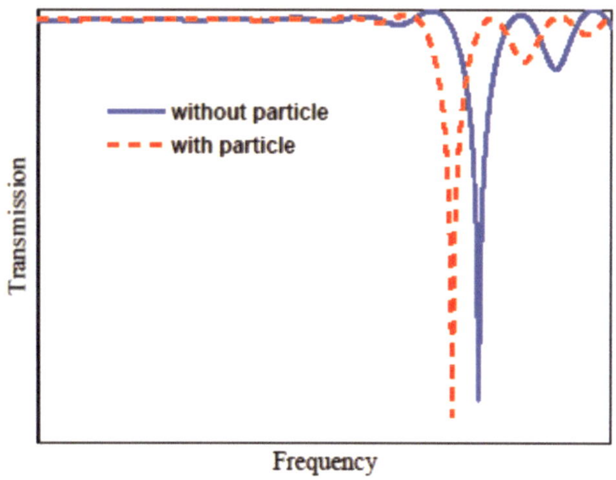

Figure 4. Shift of the resonance frequency due to attachment of particle [17].

2.2.1. Shift of resonance frequency versus size of attached nanoparticle

Assuming the conserved intensity of resonant light and smallness of the attached particle compared to the wavelength, the frequency of each resonant photon shifts by$\Delta\omega_r$ according to [15] and [18].

$$\hbar\,\Delta\omega_r \cong -\frac{\alpha_{ex}}{2}\left\langle E^2(r_v,t)\right\rangle \tag{18}$$

where $\langle E^2(r_v, t)\rangle$ is the time average of the square electric field at the nanoparticle position (r_v). This field is generated due to a single photon resonant state. $\alpha_{ex}$is the isotropic excess polarizability including local field effects [11]. By manipulations, the fractional frequency shift is obtained as [15]

$$\frac{\Delta\omega_r}{\omega_r} \cong -\frac{(\alpha_{ex}/\varepsilon_0)|E_0(r_v)|^2}{2\int\varepsilon_r(r)|E_0(r)|^2\,dV} \tag{19}$$

where E_0 is the electric field amplitude, and $\varepsilon_r(r)$ is the dielectric constant throughout the resonator. The maximum shift for a nanoparticle of radius size a_v absorbing on the equator is obtained as

$$\left(\frac{\Delta\lambda_r}{\lambda_r}\right)_{\max} \cong D\frac{a_v^3}{R^{5/2}\lambda_r^{1/2}}e^{-a_v/L} \tag{20}$$

where L is the characteristic length of the evanescent field, and D is dimensionless dielectric factor associated with both the microsphere and nanoparticle. L and D are given by

$$L \approx (\lambda/4\pi)\left(n_s^2 - n_m^2\right)^{-1/2} \tag{21}$$

$$D = 2n_m^2(2n_s)^{1/2}\left(n_{np}^2 - n_m^2\right)/\left(n_s^2 - n_m^2\right)\left(n_{np}^2 + 2n_m^2\right) \tag{22}$$

where n_s ,n_m and n_{np} are the refractive indices of the microsphere, surrounding medium and nanoparticle respectively [11]. The shifts of resonance frequency for three radiuses of particle, 100nm, 150nm, and 200nm have been obtained through FEM simulation. Comparison between simulation results and calculated results extracted from Eq. (20) has been provided in Fig 5. There is good agreement between our simulation results and Eq. (20). The amount of resonance frequency shift increases by enhancement of particle's radius. This fact is reasonable because the path of light through the attached particle increases further when the radius of particle is enhanced. It is noteworthy to mention that the size, shape and refractive index of particles for simulation are selected regarding the characteristics of viruses. Majority of viruses are spherical in shape and their radius size are between 20 to 400 nm [19].

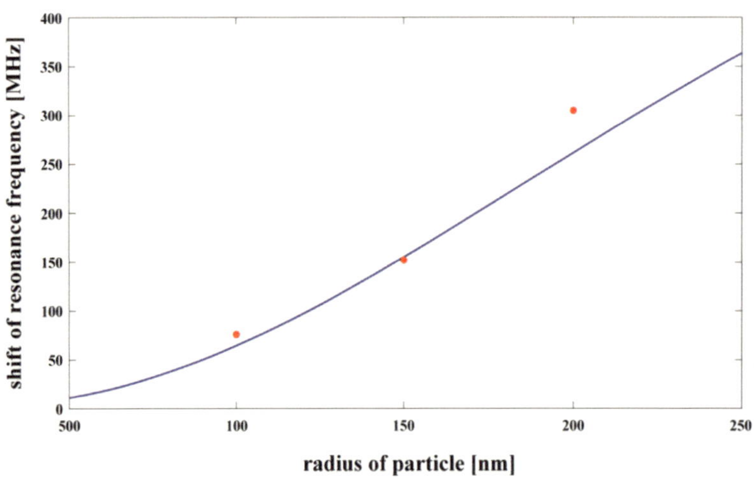

Figure 5. Shift of resonance frequency versus radius of particles [17].

3. Fiber optic sensors

One type of optical sensors is fiber optic sensors which have exclusive advantages such as low weight, capability of self-referencing and serial fashion multiplexing. According to market research report, worldwide use of fiber optic sensors will reach $3.39 billion in 2016 [20]. In the following diagram increasing trend of fiber optic sensors market is shown.

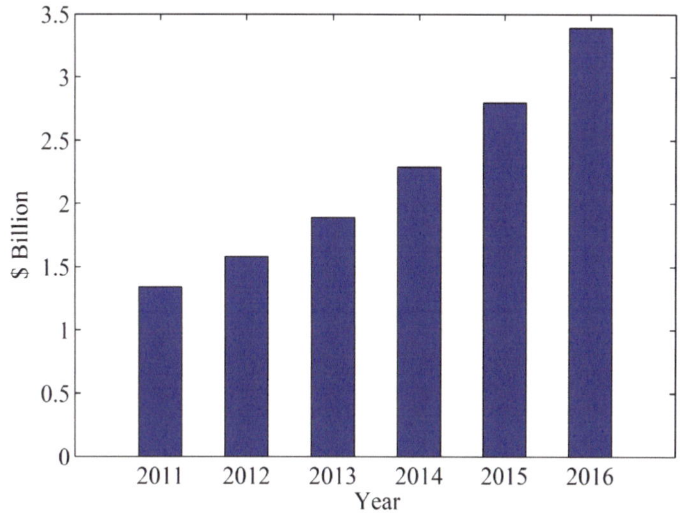

Figure 6. Increasing trend of market of fiber optic sensors.

Fiber optic sensors are classified in two categories: extrinsic fiber optic sensors and intrinsic fiber optic sensors. Extrinsic fiber optic sensors are optical fibers that lead up to and out of a "black box" that modulates the light beam passing through it in response to an environmental effect [21]. Intrinsic fiber optic sensors rely on the light beam propagating through the optical fiber being modulated by the environmental effect either directly or through environmentally induced optical path length changes in the fiber itself [21]. In the following, several fiber optic sensors which are utilized for surgeries have been investigated. Theses fiber optic sensors are intrinsic type.

3.1. Smart surgical instruments based on fiber optic sensors

Nowadays, minimally invasive surgeries are becoming more common as an efficient type of operations in hospitals. By minimally invasive surgeries, surgeon can access to internal body organs of patient through small incisions instead of large openings. During this type of surgeries, surgeons use related specific instruments like clip applicators, laparoscopic graspers, and catheters. Utilizing minimally invasive surgeries decreases recovery time and discomfort for patient. Despite all advantages of minimally invasive surgeries, it has some drawbacks. In open surgery, accuracy of operation is high due to direct visual and touch ability of the surgeon. On the other hand, in minimally invasive surgery, surgeon loses tactile feedback. Furthermore, navigation of minimally invasive instruments is a challenge in minimally invasive surgeries. Smart minimally invasive surgical instruments have been introduced to overcome such drawbacks. Catheter ablation surgery is a minimally invasive surgery which is performed by surgeon in order to cure cardiac arrhythmias. In this MIS surgery, specific catheters are inserted into inside of heart through main blood vessels of body (Fig. 7, Fig. 8). Then, part of heart which causes cardiac arrhythmias is ablated by ablation catheter [22].

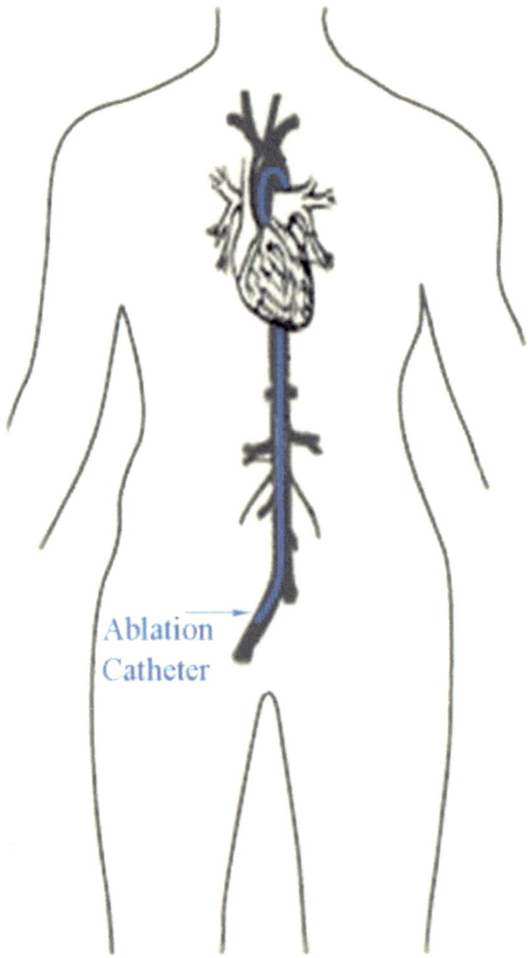

Figure 7. Catheter is inserted into inside of heart through Aorta artery.

http://dx.doi.org/10.5772/57465

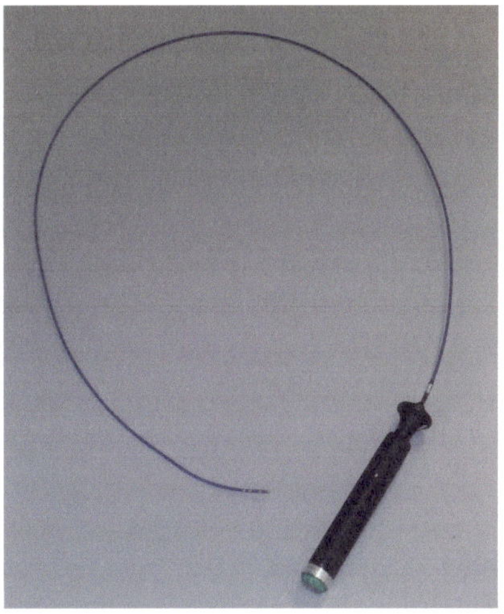

Figure 8. RF ablation catheter.

Outcome of this surgery highly depends on force applied by the ablation catheter's tip to heart tissue. Insufficient force could leads to incomplete ablation, while excessive force may result in serious injury and even perforation [23]. Therefore, to successfully ablate live tissue, the catheter should be applied to the tissue with proper force. Ablation catheters with force sensing capability have been emerged recently in order to help surgeon in performing ablation surgery with more accuracy and safety. Many papers and patents have been published in the area of smart ablation catheters. First introduced generation of smart ablation catheters utilize force sensor in the shaft out-side the patient's body [24, 25]. They provide some information of catheter-tip contact forces, but frictional forces due to the interaction between the catheter and blood vessels disturb the measurement, so exact contact force between catheter's tip and heart tissue cannot be measured. Various types of force sensors like capacitive, inductive, resistive and optical sensors can be integrated into the ablation catheter's tip in order to measure contact force, however, fiber optic sensors have been utilized efficiently rather than other types because of their exclusive advantages. Tacticath is one of the smart ablation catheters, provided by Endosense medical technology company that measures the contact force between catheters tip and heart tissue based on fiber optic sensor [26]. Contact force between catheter's tip and heart tissue can be decomposed into three main axes, one axial force and two lateral forces. Fig. 9 shows concept of axial and lateral force with respect to catheter's tip.

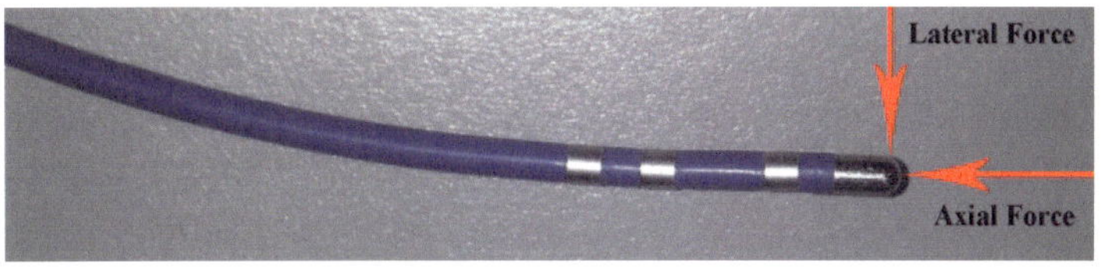

Figure 9. Axial and lateral force with respect to catheter's tip.

Smart ablation catheters that utilize fiber optic sensor as sensing element can be categorized into two main groups based on sensing technology: FBG-based and light intensity modulation based. The basic principle of operation commonly used in a FBG-based sensor is shift of Bragg wavelength. The Bragg wavelength of a FBG (Λ_B) is given by $\Lambda_B=2n\Lambda$, where n is effective index of core and Λ is grating period. Shift of Bragg wavelength ($\Delta\Lambda_B$) due to strain (ε) can be expressed by below expression [27] :

$$\Delta\lambda_B = 2n\Lambda\left(1-(\frac{n^2}{2})(P_{12}-\nu(P_{11}+P_{12}))\right)\varepsilon \tag{23}$$

where P_{11} and P_{12} are strain-optic coefficients and ν is Poisson's ratio.

FBGs can be used as strain gauges in distal end of the catheter to measure contact force between catheter's tip and heart tissue. Ref. [28] suggests a design which utilizes groups of axially oriented strain gauges (for example FBGs) embedded on distal end of minimally invasive surgical instrument (for instance ablation catheters) to measure forces at the distal end of instrument (Fig. 10). This sensor can measure lateral applied force without temperature sensitivity but temperature variation makes error in axial force measurement for this sensor.

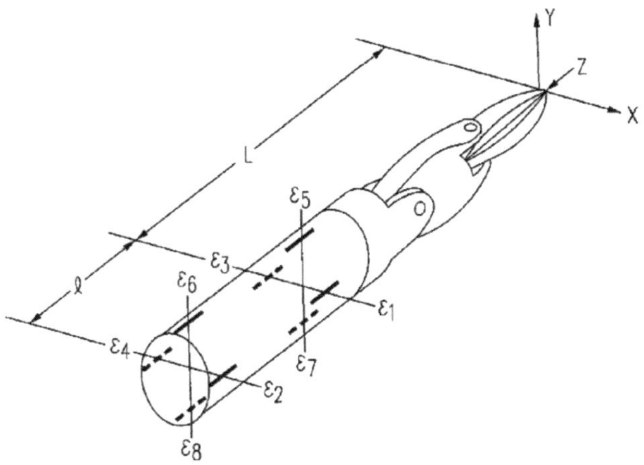

Figure 10. A minimally invasive surgical instrument with 8 embedded strain gauge for tri-axial contact force sensing (Reference [28]).

Four optical fibers, that each one has two serial FBGs, can be utilized as embedded strain gauges illustrated in Fig. 10. In such a smart minimally invasive instrument, lateral forces are calculated by $F_x=(\varepsilon_1-\varepsilon_2-\varepsilon_3+\varepsilon_4)EI/2lr$ and $F_y=(\varepsilon_5-\varepsilon_6-\varepsilon_7+\varepsilon_8)EI/2lr$. Here, E is the Young's modulus of material, I is the section moment of inertia, l is the distance between two inline strain gauges, and r is the radius from z-axis to the acting plane of the gauges. Axial force is calculated by $F_z=(\varepsilon_1+\varepsilon_2+\varepsilon_3+\varepsilon_4+\varepsilon_5+\varepsilon_6+\varepsilon_7+\varepsilon_8)EA/8$. Here, A is cross-sectional area of instrument. Calculated lateral forces (F_x and F_y) are temperature insensitive because of related

http://dx.doi.org/10.5772/57465

differential relation. Light intensity modulation can be used for contact force detection. In this method, distance between a reflector plane and tip of an optical fiber change due to applied force. This change of distance results in change of intensity of reflected light. Fig. 11 shows a smart ablation catheter with contact force sensing capability. Fig. 4 shows details of this catheter. An optical fiber, embedded inside of the catheter, emits light to surface of reflector plane. Intensity of reflected light, which is received by the same optical fiber, depends on distance between optical fiber's tip and reflector plane. When a contact force is applied, the elastic material (which is a polychloroprene rubber) deforms, decreasing the distance between optical fiber's tip and reflector, subsequently, increasing intensity of reflected light. It is noteworthy to mention that a 2 x 1 optical coupler is used in order to separate emitted light from reflected light.

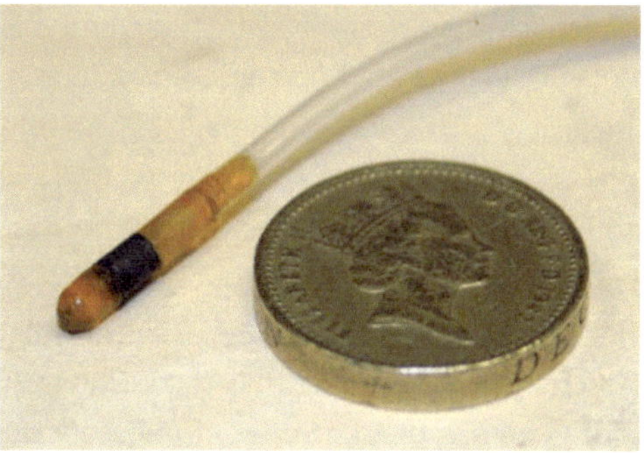

Figure 11. Ablation catheter with force sensor (Reference [29]). The outer diameter is 3mm (9Fr in French scale which is used to express size of catheters).

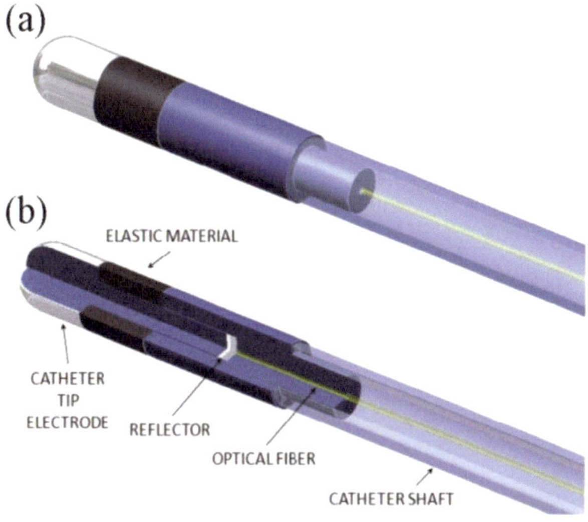

Figure 12. (a) Ablation catheter with force sensor. (b) Detailed view of this ablation catheter (Reference [29]).

Components of this sensor were created rapid prototyping technology and are made of photoractive acrylate polymers. Intensity of reflected light and, consequently, the voltage output produced from an optoelectronic circuit are nonlinear functions of the distance between optical fiber end face and reflector surface [30]. Voltage output that can be detected from an ideal optoelectronic circuit is given by:

$$V = \frac{\pi}{2}\sigma_r k_v w^2 I_0 (1 - e^{\frac{-d^2}{2w^2}}) \tag{24}$$

In above equation, σ_r is a parameter that represents light losses due to bending and misalignment of optical fiber, k_v is conversion factor that related light flux to voltage output, I_0 is maximum light intensity of reflected light received by fiber, and d is diameter of optical fiber. Also, in above equation, w is defined as $w = 2.h\,.\tan(\gamma) + w_0$. Here, h is distance between fiber end surface and reflector surface, γ is divergence angle, w_0 is mode-field radius related to light intensity distribution profile of transmitted light beam. As it can be inferred from equation (24), bending of optical fiber affects voltage output. This is due to bending losses which decrease intensity of transmitted light. During ablation surgery, catheter has several bending, thus an error occurs in contact force measurement due to bending losses. This is a disadvantage of light intensity modulation method of sensing in comparison with FBG method of sensing. Fig. 13 shows voltage output versus applied force to aforementioned catheter's tip.

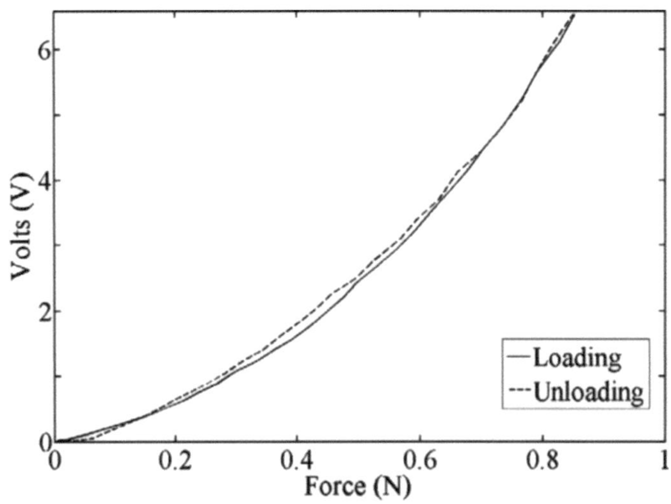

Figure 13. Loading and unloading hysteresis cycle of catheter force sensor.

Fig. 13 shows that this force sensor embedded in ablation catheter can operate with very low hysteresis (maximum calculated hysteresis ratio=2.35%). Below quadratic equation can be used to describe nonlinear behavior of force sensor:

http://dx.doi.org/10.5772/57465

$$V = AF^2 + BF \tag{25}$$

where V is the voltage output of the force sensor, when a force F is applied at the catheter tip, and A, B are the calibration coefficients. It is noteworthy to mention that this smart catheter can measure only applied axial forces. For tri-axial contact force sensing, several optical fibers should be utilized. Fig.14 shows a tri-axial ablation catheter force sensor [31]. This sensor is able to measure axial forces of up to 0.50N and lateral forces up to 0.45N with force sensing resolution of 0.01N.

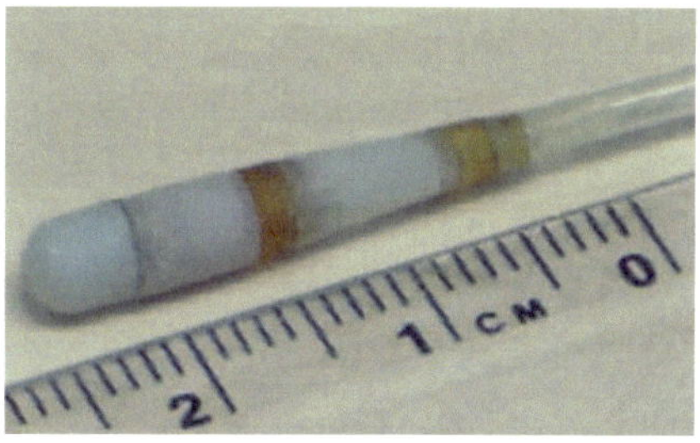

Figure 14. Ablation catheter with tri-axial force sensor

Fig.15 depicts detailed view of this smart ablation catheter. This smart catheter has four optical fibers. Three of them are utilized for tri-axial contact force measurement and fourth is reference optical fiber in order to eliminate of bending caused error. The reference optical fiber is always in contact with related reference reflector and, hence, continuously recording the maximum light intensity that can be received by the fibers. Thus, light fluctuations that are not due to contact force between catheter's tip and heart tissue are detectable.

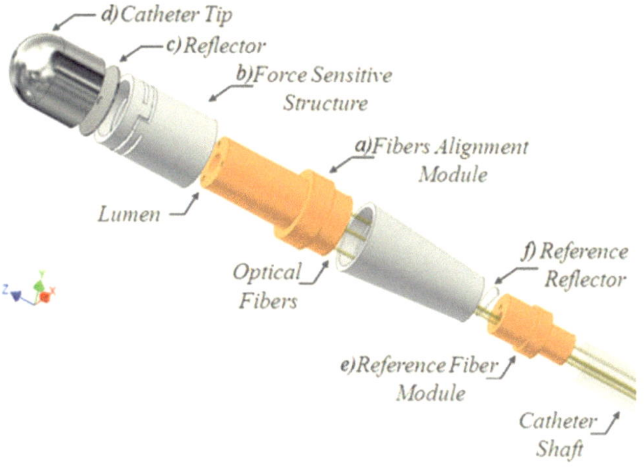

Figure 15. Exploded view of ablation catheter with tri-axial force sensor.

High-resolution rapid prototyping technique was employed to create sensing part of this smart catheter using polymeric material. Fig.16 shows comparison results between manufactured smart catheter and standard force sensor (ATI-Nano 17). Both force signals behave similarly with an rms error of 0.03N for the axial case and 0.021N for the lateral case, resulting in a sensor accuracy of 6%.

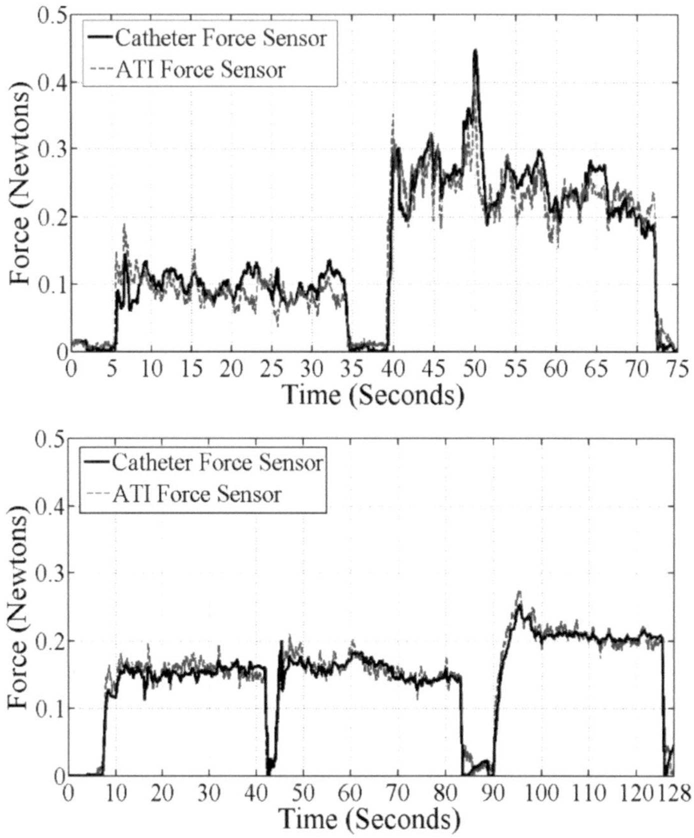

Figure 16. Comparison of the force signals obtained from smart ablation catheter and the standard force sensor, upper panel is related to axial force and lower panel is related to lateral force.

In order to increase the accuracy of minimally invasive surgeries, robotically operation of MIS have been introduces and utilized. The instruments are mounted on robot manipulators controlled by surgeon through joysticks [32-33]. Robots need feedbacks from mounted smart minimally invasive surgeries instruments in order to navigate, control and operate them. Magellan and Sensei-x are examples of robotic catheter systems, manufactured by Hansen Medical [33]. They can also give three dimensional figure of catheters shape inside of patient body. Shape sensing of catheters are based on FBG bending sensors [34].

The amount of bending in a structure is determined by a parameter namely K and it is defined as:

$$K = \frac{1}{R} \tag{26}$$

The unit of K is [m^{-1}]. Bent structure can be supposed on the sector of a circle. In relation (26), R with unit [m] is the radius for this imaginary circle. Fig. 17 depicts this matter.

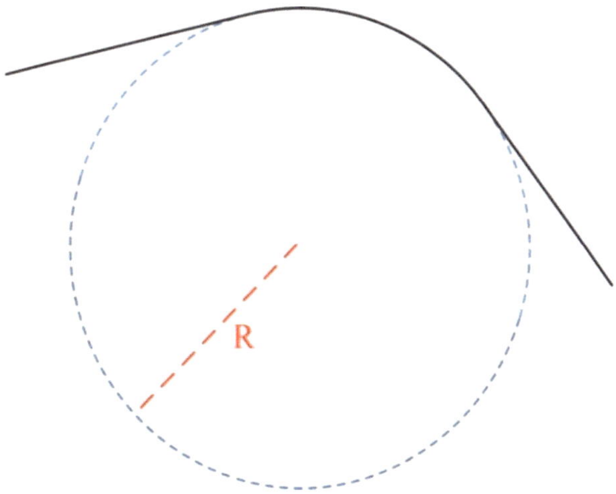

Figure 17. A bent structure on the sector of an imaginary circle with radius R

The output of a bent sensor is the amount of structure's bending.

If a catheter get bent around an imaginary circle with radius of R, like Fig. 18, engendered strain in this fiber is [35] :

$$\varepsilon_z = Kx \tag{27}$$

Engendered strain is extension for x>0 and compression for x<0.

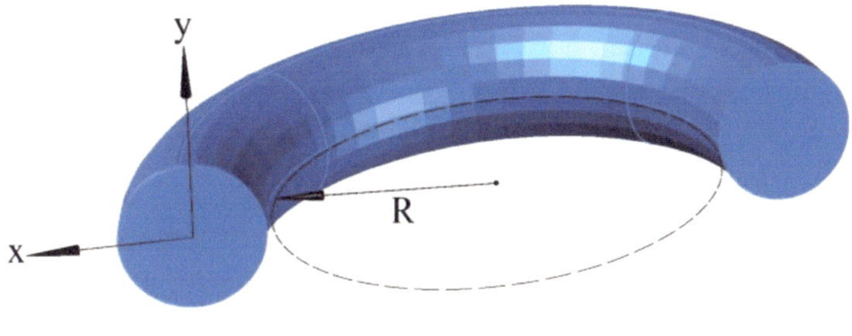

Figure 18. Assumed coordinates of fiber and radius of bending

Optical fiber which is placed inside catheter in eccentric position, can measure the amount of bending [36]. Related smart catheters have several optical fibers. These optical fibers have several serial FBGs that measure bending and its direction in different parts of catheter. Assembling obtained shape of bending related to different parts, it is possible to extract the shape of catheter. Three optical fibers can be utilized in order to extract the shape of the catheter [37].

4. Summary

Optical sensors are growing fast in all technological fields. This type of sensors is utilized in biomedical engineering, efficiently and according to most roadmaps future is well brilliant for optical sensors in this field of science and technology. Many corporations have been emerged to invest in research, development and fabrication of instruments with optical sensing capability. In this chapter important types of optical sensors including WGM-based sensors and fiber optic sensors, have been investigated with considering their applications in biomedical field.

Author details

A. Rostami*, H. Ahmadi, H. Heidarzadeh and A. Taghipour

*Address all correspondence to: rostami@tabrizu.ac.ir

School of Engineering-Emerging Technologies, University of Tabriz, Tabriz, Iran

References

[1] S. Q. Lud, M. G. Nikolaides, I. Haase, M. Fischer and A.R. Bausch, "Field Effect of Screened Charges: Electrical Detection of Peptides and Proteins by a Thin Film Resistor," Chem. Phys. Chem. Vol. 7, no. 2, pp. 379-384.

[2] M. I. R. Gaso, C. M. Iborra, A. M. Baides, and A. A. Vives, "Surface Generated Acoustic Wave Biosensors for the Detection of Pathogens: A Review," Sensors, vol. 9, pp. 5740- 5769, 2009.

[3] A. M. Armani, R. P. Kulkarni, S. E. Fraser, R. C. Flagan, and K. J. Vahala, "Label-Free, Single-Molecule Detection with Optical Microcavities", Science, vol. 317, pp. 783-787, 2007.

[4] H. Zhu, I. M. White, J. D. Sutter, M. Zourob, and X. Fan, "Miniaturized opto-fluidic ring resonator for sensitive label-free viral detection" Proc. of SPIE, vol. 6896, 2008.

[5] K. J. Vahala, "Optical Microcavities," Nature, vol. 424, pp. 839-846, 2003.

[6] J. P. Laine, Design and Applications of Optical Microsphere Resonators," Helsinki University of Technology, Materials Physics Laboratory, 2003.

[7] L. Collot et al., "Very high-Q whispering-gallery mode resonances observed on fused silica microspheres", Europhys. Lett. Vol. 23, pp. 327-334, 1993.

[8] M. L. Gorodetsky et al., "Optical microsphere resonators: optimal coupling and the ultimate Q," SPIE, vol. 3267, pp. 251-262, 1998.

[9] B. E. Little, J. P. Laine, H. A. Haus, "Analytic Thory of Coupling from Tapered Fibers and Half-Blocks into Microsphere Resonators," IEEE Journal of Lightwave Technology, vol. 17, no. 4, pp. 704-715, 1999.

[10] J. Brockman, The Next Fifty Years, NewYork: Vintage Books, 2002, pp 289-301.

[11] F. Vollmer, S. Arnold, and D. Keng, "Single Virus Detection from the Reactive Shift of a Whispring-gallery Mode," PNAS, vol. 105, no. 52, 2008.

[12] H. C. Ren, F. Vollmer, S. Arnold, and A. Libchaber, "High-Q microsphere biosensor-analysis for adsorption of rodlike bacteria," Optics Express, vol. 15, no. 25, 2007.

[13] F. Vollmer, and S. Arnold, "Whispering-gallery-mode biosensing: label-free detection down to single molecules," Nature Methods, vol. 5, no. 7, 2008.

[14] F. Vollmer, and S. Arnold, "Whispering-gallery-mode biosensing: label-free detection down to single molecules," Nature Methods, vol. 5, no. 7, 2008.

[15] A. Arnold, M. Khoshsima, and I. Teraoka, "Shift of whispering-gallery modes in microsphere by protein adsorption," Optics Letters, vol. 28, no. 4, pp.272-274, 2003.

[16] M. L. Gorodetsky, A. A. Savchenkov, and V. S. Ilchenko, "Ultimate Q of optical microsphere resonators," Optics Letters, vol. 21, no. 7, pp.453-455, 1996

[17] H. Ahmadi, H. Heidarzadeh, A. Taghipour, A. Rostami, H. Baghban, M. Dolatyari and G. Rostami, Evaluation of Single Viruse Detection through Optical Biosensor Based on Microsphere Resonator submitted article to fiber and integrated optics journal

[18] Arnold S, Ramjit R, Keng D, Kolchenko V, Teraoka I (2008) Microparticle photophysics illuminates viral biosensing. Faraday Discuss 137:65–85.

[19] E. Solomon, L. Berg, and D. W. Martin, Biology, Stamford, Connecticut: Cengage Learning, 2010, pp.502.

[20] Market Research Reports, Global Information Inc.

[21] S. Yin, P. B. Ruffin, and F. T. S. Yu, Fiber Optic Sensors, New York: CRC Press, 2008, pp. 2-3.

[22] S. K. S. Huang, Catheter Ablation of Cardiac Arrhythmias, Philadelphia: Elsevier Saunders, 2010.

[23] L. Di Biase et al., "Relationship between Catheter Forces, Lesion Characteristics, "Popping," and Char Formation: Experience with Robotic Navigation System," Journal of Cardiovasc. Electrophysiol. vol. 20, no. 4, pp. 436-440, 2008.

[24] J. Rosen et al. "Surgeon-Tool Force/Torque Signatures - Evaluation of Surgical Skills in Minimally Invasive Surgery," In Proc. of Medicine Meets Virtual Reality, San Francisco, 1999.

[25] P. Kanagaratnam et al., "Experience of robotic catheter ablation in humans using a novel remotely steerable catheter sheath," Journal of Interv. Card. Electrophysiol. Vol. 20, pp. 19-26, 2008.

[26] Endosense Inc., www.endosense.com/

[27] A. D. Kersey et al., "Fiber Grating Sensors," Journal of Lightwave Technology, Vol. 15, No. 8, 1997.

[28] S. J. Blumenkranz, D. Q. Larkin, Force and Torque Sensing for Surgical Instruments, US Patent, US20070151390, 2007.

[29] P. Polygerinos et al., "MRI-Compatible Intensity-Modulated Force Sensor for Cardiac Catheterization Procedure," IEEE Transaction on Biomedical Engineering, Vol.58, No. 3, 2011.

[30] P. Polygerinos, L. D. Seneviratne and K. Althoefer, "Modeling of Light Intensity-Modulated Fiber Optic Displacement Sensors," IEEE Transaction on Instrumentation and Measurement, Vol. 60, No. 4, 2011.

[31] P. Polygerinos et al., "Triaxial Catheter-Tip Force Sensor for MRI-Guided Cardiac Procedures," IEEE/ASME Transactions on Mechatronics, Vol. 18, No. 1, 2013.

[32] Intuitive Surgical Inc., http://www.intuitivesurgical.com/

[33] Hansen Medical Inc., http://www.hansenmedical.com/

[34] M. J. Roelle et al., Fiber Optic Instrument Sensing System, US Patent, US201110319815, 2011.

[35] F. P. Beer, E. R. Johnston and J. T. Dewolf. 2006. Mechanics of Materials. Mc Graw Hill.

[36] A. Rostami et al., "Grating-Based Fiber Bending Sensors with Wide Bending Range," In Proc. of ISOT'12 Intl. Symposium on Optomechatronic Technologies, 2012.

[37] J. P. Moore and M. D. Rogge, "Shape sensing using multi-core fiber optic cable and parametric curve solutions", Optics Express, Vol. 20, No. 3, 2012.

Chapter 6

Self-Protected Sensor System Utilizing Fiber Bragg Grating (FBG)-Based Sensors

Chien-Hung Yeh and Chi-Wai Chow

Additional information is available at the end of the chapter

http://dx.doi.org/10.5772/57417

1. Introduction

Utilization of fiber Bragg gratings (FBG) to serve as the fiber-optic sensors are important researches for the optical fiber sensing applications [1–4]. Hence, when the strain or temperature variations are imposed on the FBG-based sensor, the Bragg wavelength will drift due to the index change of the FBG, and thus the detected wavelength will also shift. And so, according to the wavelength shift phenomenon of FBG, we could easily observe the strain and temperature variations for sensing [3]. In the distributed fiber sensors of the intelligent sensing network system, the FBG-based sensor has been studied and identified as an important sensing element for sensing the change of temperature and strain [1–6]. Hence, the large-scale FBG sensor system could be built and achieved easily by wavelength multiplexing technology [7–9].

During the strain sensing, if the payload applied on the FBG is over its limitation, FBG breakage will occur. Furthermore, the fiber-fault of FBG-based sensing network also could affect the survivability and reliability. As a result, how to improve and enhance the reliability and survivability of sensor network becomes the essential issue for further study. To keep the survivability of an FBG sensor system against fiber fault, a few self-protection schemes have been reported [10,11]. However, these methods required optical switch (OSW) in each remote node (RN) to re-route the sensing path for detecting and protecting the fiber fault. It was hard to control and judge the direction of OSW in each RN and also increased the complexity and cost of FBG-sensor networks. In addition, even though these FBG sensor systems were passive sensing networks in multi-ring schemes in the past studies [1, 2]. But, these ring sensor architectures required many numbers of RNs, consisting of optical coupler (OCP), to produce multi-ring configurations and result in the huge fiber connect points to reduce the power budget after run trip transmission.

In this work, we will introduce two different FBG-based sensor networks utilizing the passive system design. In the first design, a self-protected passive FBG sensor network, which enhances the reliability and survivability in long-reach fiber distance, is proposed and experimentally investigated. The sensing mechanism is based on a 25 km cavity length erbium-doped fiber (EDF) ring laser for detecting the multiple FBG sensors in the network. And we only use an optical coupler (OCP) on RN to produce multi-ring configuration. The advantage of the proposed fiber laser scheme for detecting and monitoring the FBG sensors in the long distance self-protected scheme can facilitate highly reliable and survivable operation. Besides, the long distance sensing system can be integrated in fiber access network to reduce the cost of sensor infrastructure in the future. In the second design, we propose and demonstrate a multi-ring passive sensing architecture, which does not have any active components in the entire network. In this experiment, the network survivability and capacity for the multi-point sensor systems are also enhanced. Moreover, the tunable laser source (TLS) is adopted in central office (CO) for FBG sensing. The survivability of an eight-point FBG sensor is examined and analyzed. Due to the passive sensor network, the cost-effective and intelligent sensing system is entirely centralized by the CO. As a result, the experimental results show that the proposed system can assist the reliable FBG sensing network for a large-scale and multi-point architecture.

2. First FBG sensor network architecture

The proposed LR self-healing FBG-based sensor system was consisted of a central office (CO) and multiple sensing networks, as shown in Figure 1. Here, two 25 km long single-mode fibers (SMFs) were used to connect to the CO and a N×N optical coupler (OCP), which was located at remote node (RN). The upper (path "1") and lower SMFs (path "2") were used to serve as the feeder fibers for sensing signal transmission. The N×N OCP could produce N×1 fiber ring architectures. And each ring scheme could use M FBGs for system sensing. A wavelength-tunable laser (WTL) source was used to detect and monitor the FBG sensors in the CO. For the initial deployment of the sensing network, the power budget (i.e. the power of the laser source and the split-ratio of the sensing network) should be carefully considered. If the power budget is not enough due to high split ratio, fiber amplifiers between CO and RN can be used to solve this issue. Besides, according to current passive access network standards [12], this proposed 25 km long sensor system can be also integrated in the fiber access system to enhance the use of capacity fiber and reduce the cost of sensor infrastructure.

To realize and perform the proposed FBG-based sensor network, the experimental setup was performed on a simplified version as illustrated in Figure 2. The CO was constructed by a WTL, a 1×2 OSW and an optical spectrum analyzer (OSA). In this measurement, the WTL was consisted of an erbium-doped fiber amplifier (EDFA), a tunable bandpass filter (TBF), a 4×4 OCP, a polarization controller (PC), a variable optical attenuator (VOA). The 980 nm pumping laser of EDFA operated at 215 mA and the saturated output power of the EDFA was around 16 dBm at 1530 nm. The tuning range, insertion loss and 3 dB bandwidth of TBF used was 36 nm (1526 to 1562 nm), <0.7 dB and 0.3 nm, respectively. The PC and VOA were used to control and adjust the polarization status and maintain the maximum output power. Therefore, the

http://dx.doi.org/10.5772/57417

4×4 OCP would introduce three ring sensing networks. And the two feeder fibers were 25 km long. In the experiment, the sensor network has eight FBGs, which could be used for the strain and temperature sensing applications. Moreover, the Bragg wavelengths of these FBGs, which have different reflectivity, were 1527.6 (λ_{11}), 1528.9 (λ_{12}), 1532.9 (λ_{21}), 1536.7 (λ_{22}), 1538.4 (λ_{23}), 1541.7 (λ_{31}), 1546.0 (λ_{32}) and 1556.0 nm (λ_{33}), respectively.

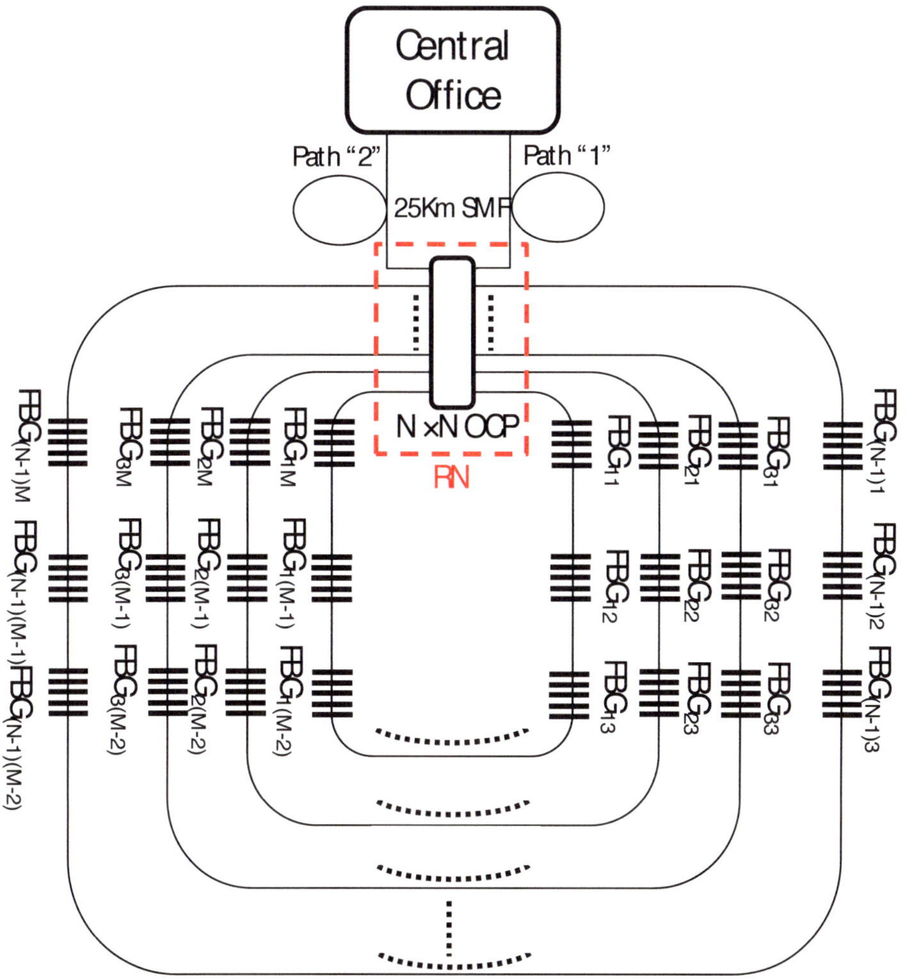

Figure 1. Proposed long distance FBG sensor system in passive multi-ring architecture. SMF: single-mode fiber; OCP: optical coupler; FBG: fiber Bragg grating; RN: remote node.

The proposed WTL scheme in CO contained the C-band erbium-doped gain operation. And the passband of the TBF, using inside gain cavity, was scanned to match the corresponding Bragg wavelength of FBG. Each FBG element served as the reflected sensor head and was connected as the part of cavity via a 25 km long fiber. Due to the inclusion of the TBF within the cavity loop, the lasing wavelength would be generated only when the filter passband was aligned in accordance with one of the FBG elements. In normal operation, when the TBF was set at one of the Bragg wavelength of the FBG, a long cavity fiber ring laser will be formed, and a lasing wavelength will be detected at the OSA located in the CO, with a 0.05 nm resolution.

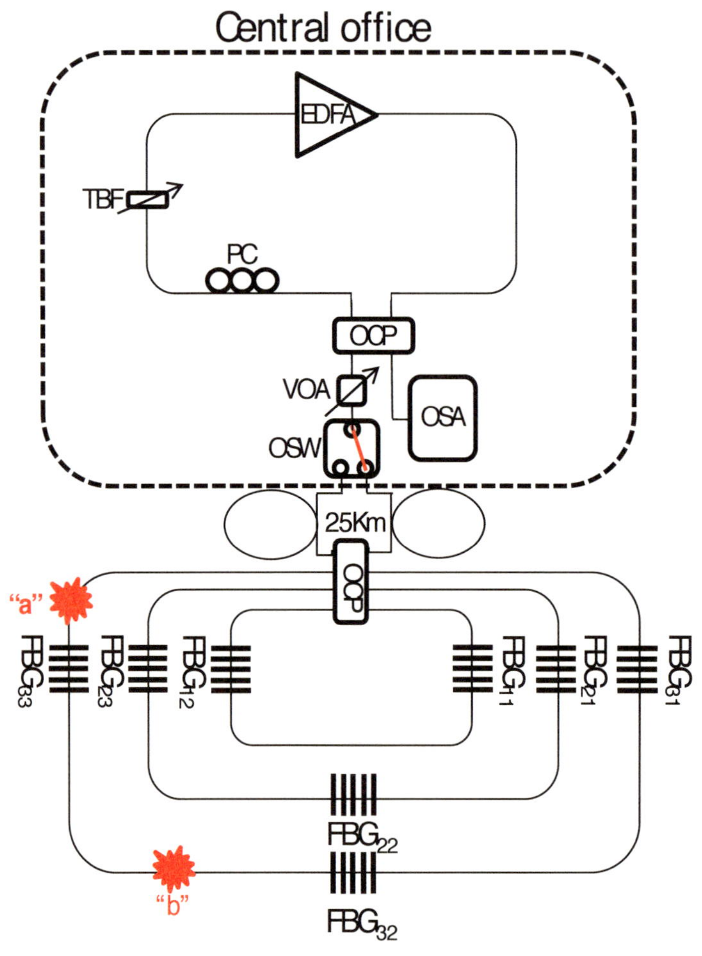

Figure 2. Experimental setup of proposed long distance self-healing FBG-based sensor system.

Initially, the OSW of CO was located on position "1" to connect to the sensor network via the upper feeder fiber (path "1") for detecting and monitoring the eight FBG sensors used, as illustrated in Figure 2. Here, Figure 3(a) shows the reflected wavelengths of the proposed FBG-based sensor network from 1527.6 to 1556.0 nm (λ_{11} to λ_{33}) via upper fiber. Here, we can observe eight lasing wavelengths in Figure 3(a) using the proposed WTL scheme during the TBF scanning. Of course, we could also connect the FBG sensor network for via the lower feeder fiber (path "2"), when the 1×2 OSW was switched to position "2", as shown in Figure 2. Hence,

Figure 3(b) presents the retrieved output wavelengths of the eight FBG sensors system via the lower feeder fiber. As shown in Figure 3(a) and Figure 3(b), no matter from which fibers to transmit the detecting signal to monitor FBG sensors, the measured reflective wavelengths are the same. However, the obtained wavelength λ_{11} of Figure 3(b) is slightly drifted due to the different forward and backward reflected spectrum of FBG_{11}. As a result, we can switch the direction of OSW for connecting the upper or lower feeder fibers to monitor the status of FBG sensors and each ring fiber simultaneously.

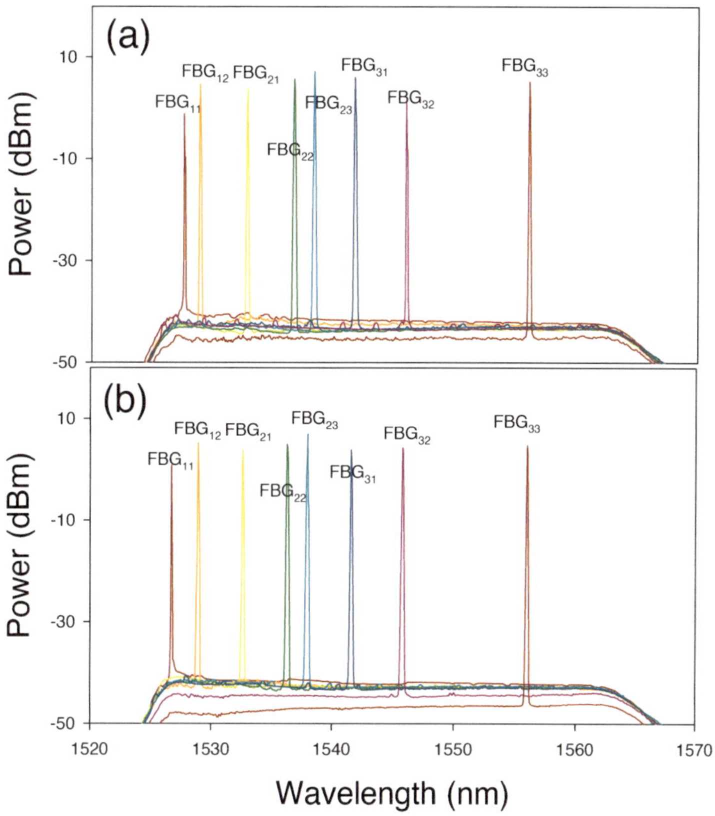

Figure 3. Output wavelengths of the proposed FBG-based sensor system from 1527.6 to 1556.0 nm via the (a) working and (b) protecting fiber links.

In the first scenario, the fiber fault may happen inside the fiber ring network. Hence, if a fiber cut occurs at point "b", as seen in Figure 2, the lasing wavelengths of λ_{32} and λ_{33} can be not measured via the upper feeder path, as shown in Figure 4(a). At this moment, the OSW will switch to position "2" to connect the lower feeder fiber for rescanning the FBG sensors. Thus, only the wavelength λ_{31} can be not retrieved by the proposed WTL scheme, as illustrated in Figure 4(b). According to the measured results of Figure 4(a) and Figure 4(b), then we can locate and ensure the cut position among FBG_{31} and FBG_{32}. Besides, if two fiber cuts occur at point "a" and "b" simultaneously in the sensor network, the lasing wavelengths of λ_{31}, λ_{32} and λ_{33} via the paths "1" and λ_{31} via the path "2" can be not measured by OSA, respectively. Hence, we can ensure the faults are among points "a" and "b". Furthermore, when the external strain is applied on FBG_{31} before broking, the wavelength shifts can be observed, as illustrated in Figure 5. Here, the maximum strain shift of FBG_{31} is nearly 2.2 nm. When the temperature of FBG increases gradually, the reflected wavelength of the FBG will shift. In the proposed FBG sensing network, certain sub-ring systems could be used for the temperature sensing. Besides, some FBG sensors are used for both temperature-sensing as well as fiber-cut location detection. In the measurement, the maximum wavelength shift of FBG_{31} is ~2.2 nm when the temperature of FBG_{31} change is around 68 °C. While the strain exceeds the limitation of FBG31, the FBG sensor will be broken. In such situation, if a fault occurs on FBG_{31}, the measured lasing wavelengths in CO are the same comparing with the results of Figure 4. However, we can first observe the wavelength drift before FBG broking. As a result, we can clearly know the differences of fiber-fault and FBG-fault in the proposed passive FBG sensor network.

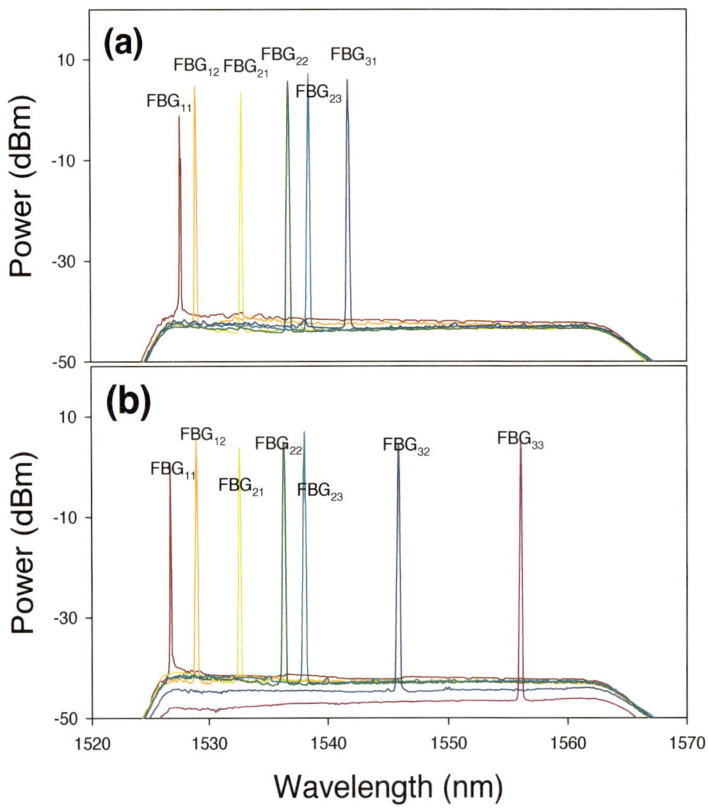

Figure 4. Output wavelengths of the proposed FBG-based sensor system via the (a) working and (b) protecting fiber links, when a cut occurs at "b" point in Figure 2.

In addition, it is necessary to ensure that each FBG sensor was in good situation in fiber sensor system. Thus, when the strain or temperature change was applied on each FBG sensor due to the environmental or artificial effect, the sensor will cause the Bragg wavelength shift. For the second scenario, while the payload of the FBG exceeds the limitation, the FBG would be broken. Here, the proposed ring sensor architecture also could find the position of the broken FBG. For example, if the FBG_{22} was broken in the proposed sensor network via the fiber path "1", the lasing wavelengths of λ_{22} and λ_{23} would not be detected, as shown in Figure 6(a). Hence, to detect the disappeared FBG sensors, the CO would control the OSW to reconnect to the lower feeder fiber (path "2") for sensing FBG. Thus, the measured output spectrum lacks the λ_{21} and λ_{22}, as shown in Figure 6(b). Comparing the two output spectra of Figurs 6(a) and 6(b), we can easily locate the fault is on FBG_{22}. Moreover, for example, when the measured lasing spectra of the λ_{11}, λ_{12}, λ_{21}, λ_{22} and λ_{23} via the fiber path "1", and λ_{11} and λ_{21} via the fiber path "2" respectively, are not detected by the proposed sensor system, we can assure that two sensors of FBG_{11} and FBG_{21} are broken at this time.

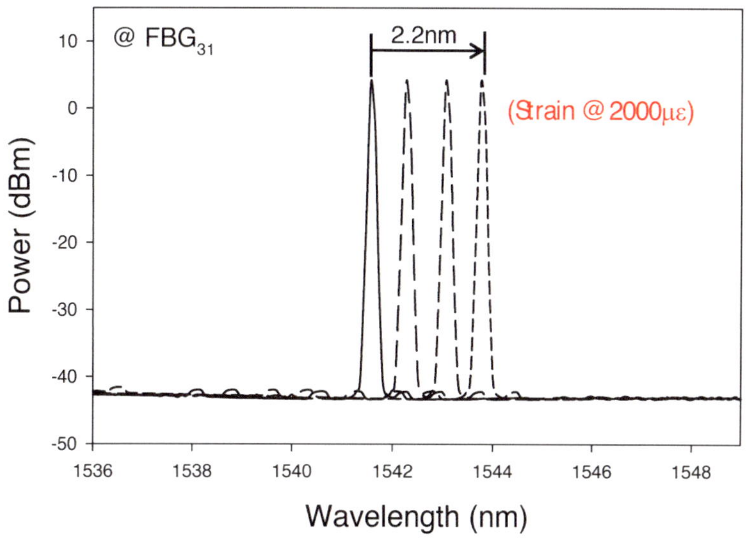

Figure 5. Reflected spectrum of FBG_{31} when the strain is applied.

We can be according to the above measured methods to ensure when the sensing path is via the feeder fiber paths "1" and "2". Initially, we can observe entire reflected wavelengths of FBG sensors via feeder path "1" when this sensing network is without FBG fault or fiber fault. When a fault is produced on FBG_{xy}, first we can observe that the reflective-wavelength of measured FBG_{xy} should be shift and then broken. Based on the measured result, we can realize the fault which is on FBG not on fiber. Besides, when the occurrence of fiber fault is produced, the all observed reflective-wavelengths are no change at this time. As a result, the proposed FBG sensor system not only can find out the fiber fault, but also detect and monitor the FBG sensors.

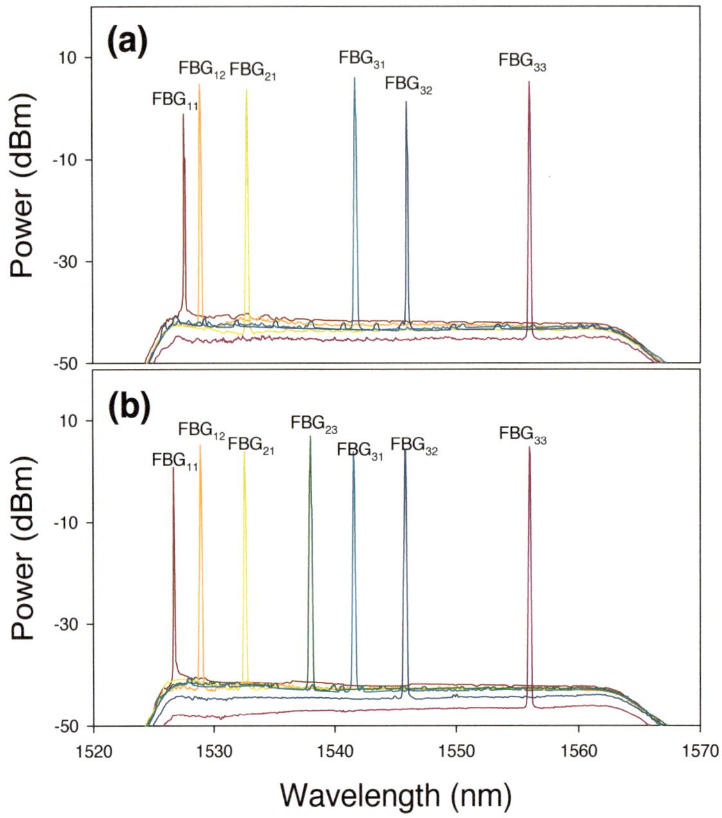

Figure 6. Measured output spectra of the sensing system via the (a) upper and (b) lower feeder fibers while a fault is on FBG_{22}.

3. Second FBG sensor network architecture

Figure 7 shows the simply self-restored ring-based architecture for passive FBG sensor system. The central office (CO) is constructed by a tunable laser source (TLS), an optical circulator (OC), a 1×2 optical switch (OS). Two output ports of OS are used to circulate a major ring. Besides, the 1×2 OS can be used to select the fiber path "a" or "b" for the sensing transmission. Assuming there are m sub-ring sensor groups in the major ring sensing system and the each groups have n FBG sensors, as illustrated in Figure 7. Each remote node (RN) uses a 2×2 optical coupler (C) to circulate sensor group and also connect next sensor group. Thus, the proposed sensing system has (m×n) FBG sensors. The sensing laser source usually uses linear-cavity erbium-doped fiber (EDF) ring laser for fiber sensor because the laser has intense output power and high OSNR [13]. In our proposed sensing network, the TLS is distributed to each RNs and delivered to each FBG sensor through each sensor groups. The TLS has the advantage of high

http://dx.doi.org/10.5772/57417

output power and optical signal to noise ratio (OSNR) to detect the FBG sensor in the network. In the sensing network, three fault locations could occur, such as in the major ring, sub-ring, and fiber sensor itself. In the following analysis, we will discuss and analyze the three situations based on the simply apparatus and ring-based architecture.

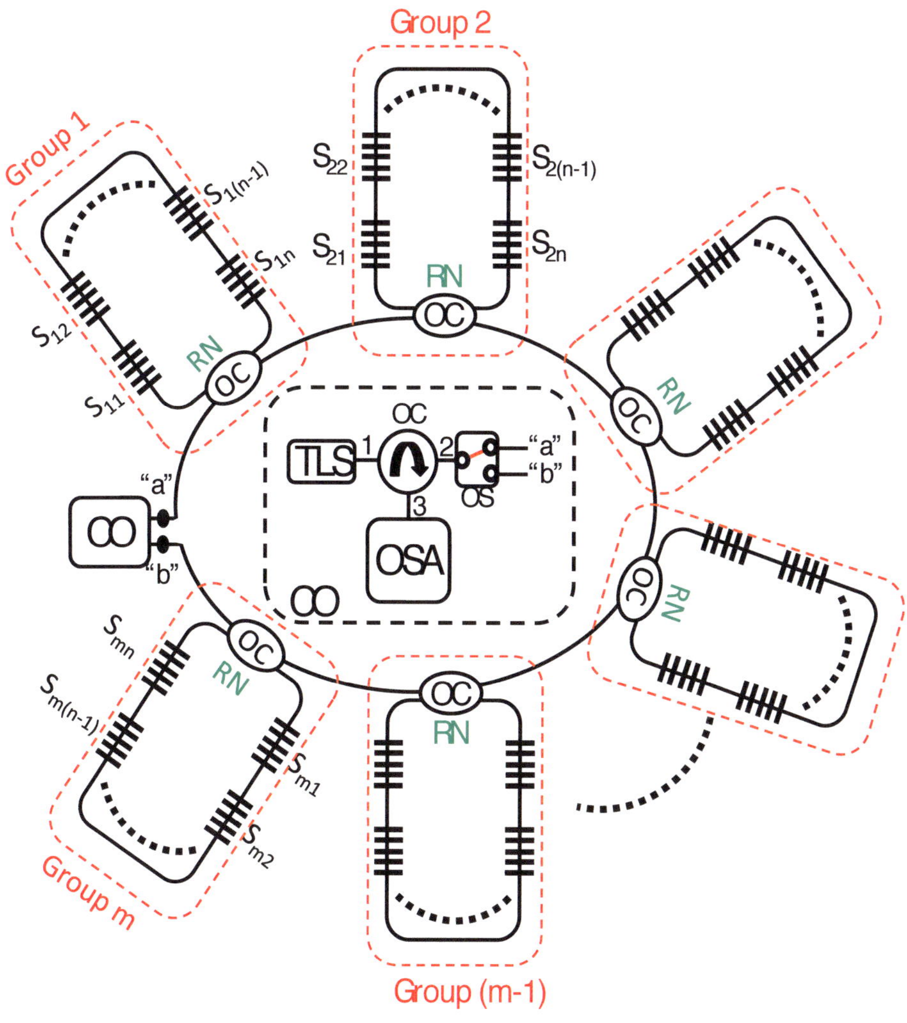

Figure 7. Proposed simply self-restored ring-based architecture for passive FBG sensor system. S: sensor, C: coupler, CO: central office, RN: remote node, OS: optical switch, OC: optical circulator, OSA: optical spectrum analyzer.

To realize and evaluate our proposed self-restored ring-based architecture for passive FBG sensor system, a simply experimental setup is shown in Figure 8. In the sensing system, the m and n of the setup are equal to four and two in the experiment. That is to say, there are eight FBG sensors (S_{mn}) using in the sensing network experiment. Each of the sensing FBGs (S_{11} to

S_{42}) are used to act as the reflected elements. In the CO, the lasing of TLS is detected by these FBGs. The Bragg wavelengths of the eight FBGs used are 1526.63, 1528.87, 1532.64, 1536.57, 1538.24, 1541.88, 1545.83 and 1555.85 nm, respectively. In addition, the fiber sensing system can also be used to accurately measure the strain and temperature perturbations applied on the FBGs. In normal status, the OS locates at point "1" to connect the path "a". Thus, the lasing wavelength from CO passes through the path "a" to detect FBG sensors. As illustrated in Figure 8(a), the green arrows show the sensing transmission path when the sensor system link without any faults. Figure 8(b) also shows the received output spectra of the eight sensing sensors from S_{11} to S_{42} by using the TLS while the sensing transmission through the path "a". When there is no fiber fault in the network, the CO would detect the eight fiber sensors in the proposed sensing system.

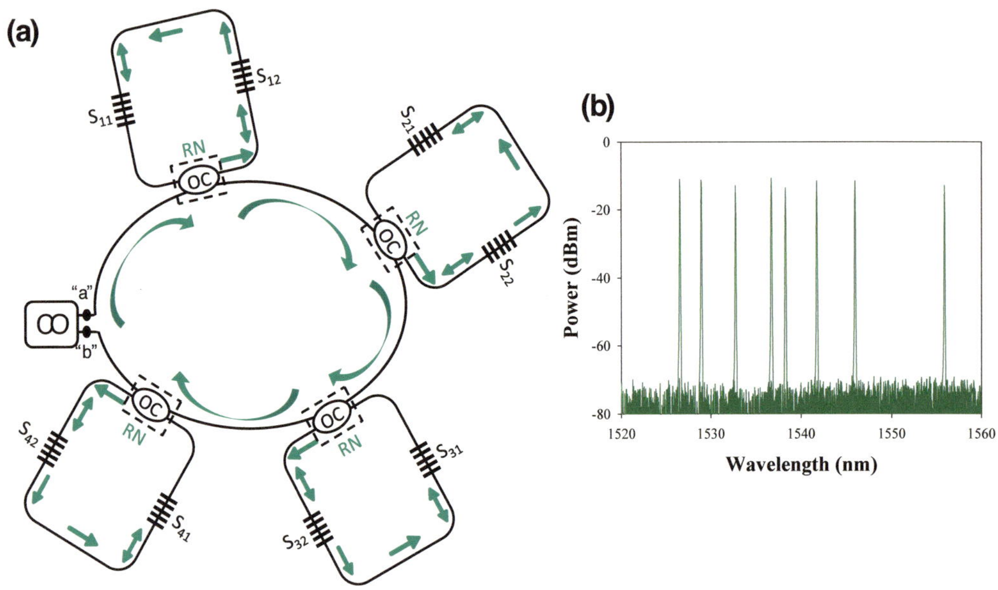

Figure 8. (a) Experimental setup for the self-restored passive FBG sensor system with eight FBS sensors from S_{11} to S_{42} without any fails. (b) Received output sensing spectra from the eight sensing FBGs in CO. (Green arrow presents the sensing transmission of the proposed network)

First, the sensor system passes through the path "a" initially for fiber sensing without any faults. When the proposed sensor network has a fiber cut on major ring between the group "2" and "3", as shown in Figure 9(a), the CO can only detect the received output sensing spectra from S_{11} to S_{22} (as seen in green line), as shown in Figure 9(b). In order to detect the residual FBG sensors, the OS could switch automatically to point "2" to link the fiber path "b". The sensing transmission is shown in red arrow, as also illustrated in Figure 9(a). When the sensing transmission is switched to path "b", the residual output sensing spectra from other sensors [as seen in red line of Figure 9(b)] can be detected. As a result, in accordance with the proposed operating mechanism, the self-restored ring-based sensor system could be protected from fiber fault on major fiber and detects the fault location approximately. If two fiber cuts are between

http://dx.doi.org/10.5772/57417

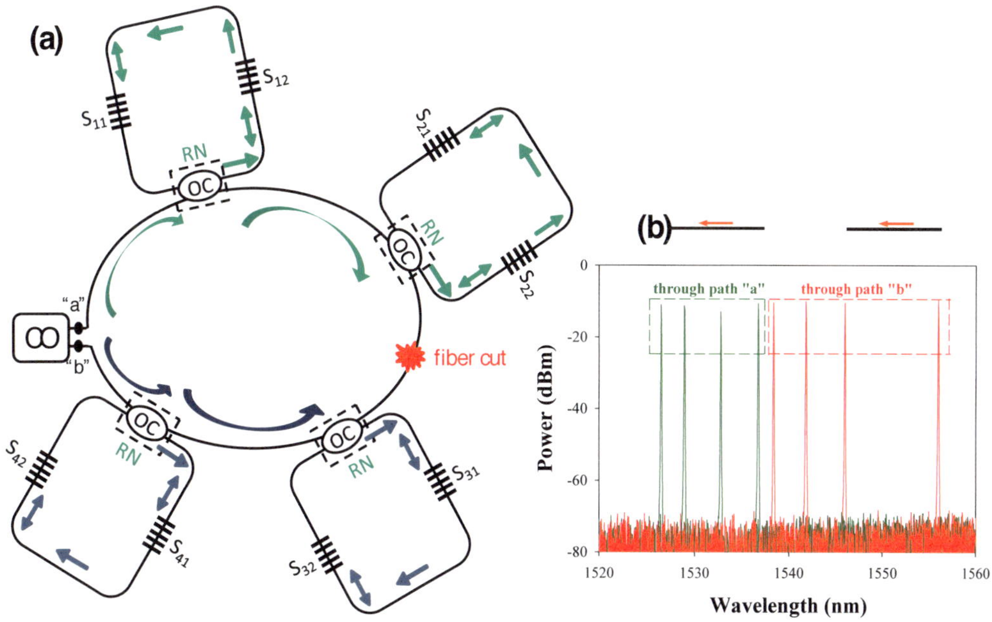

Figure 9. (a) When the sensor system has a fiber cut on major ring between the group "2" and "3" in the proposed sensing network. (b) Received output sensing spectra of the eight sensing FBGs in CO. (Green and red lines represent the sensing transmission through path "a" and "b", respectively.)

the group "2" and "3" and the group "3" and "4" in Figure 8, respectively, then the sensor S_{31} and S_{32} cannot be detected due to the two faults. However, the probability of producing two fiber faults simultaneously is very low in real sensing system.

And then, we will discuss and experiment the fiber fault in sub-ring sensor group as follows. When a fiber cut occurs between S_{21} and S_{22} in sensor group "2" through the path "a", as shown in the above of Figure 10(a), the proposed sensing network cannot detect the sensor S_{21}. Hence, as shown at the top of Figure 10(b), the output sensing spectrum lacks the sensor S_{21}. At the same time, the CO would control the OS to connect to the transmission path "b", as illustrated at the bottom of Figure 10(a). At the bottom of Figure 10(b), the output spectrum of sensor lacks the information of S_{22}. Therefore, compared with the two output sensing spectra of Figure 10(b), it can be easy to find out the fault location by detecting the FBG sensors information in the proposed network.

In sensor system, it is necessary to ensure that each fiber sensor is in good condition. Thus, when the strain or temperature change applied on FBG due to the environment or artificial effect, the fiber sensor would cause the Bragg wavelength shift. While the payload of the FBG exceeds the limitation, the FBG would cut. So, the proposed sensing system also can look for the position of broken sensor. For example, while the sensor S_{22} is broken in the proposed sensing network through the fiber path "a", the sensing network cannot detect the sensor S_{22}

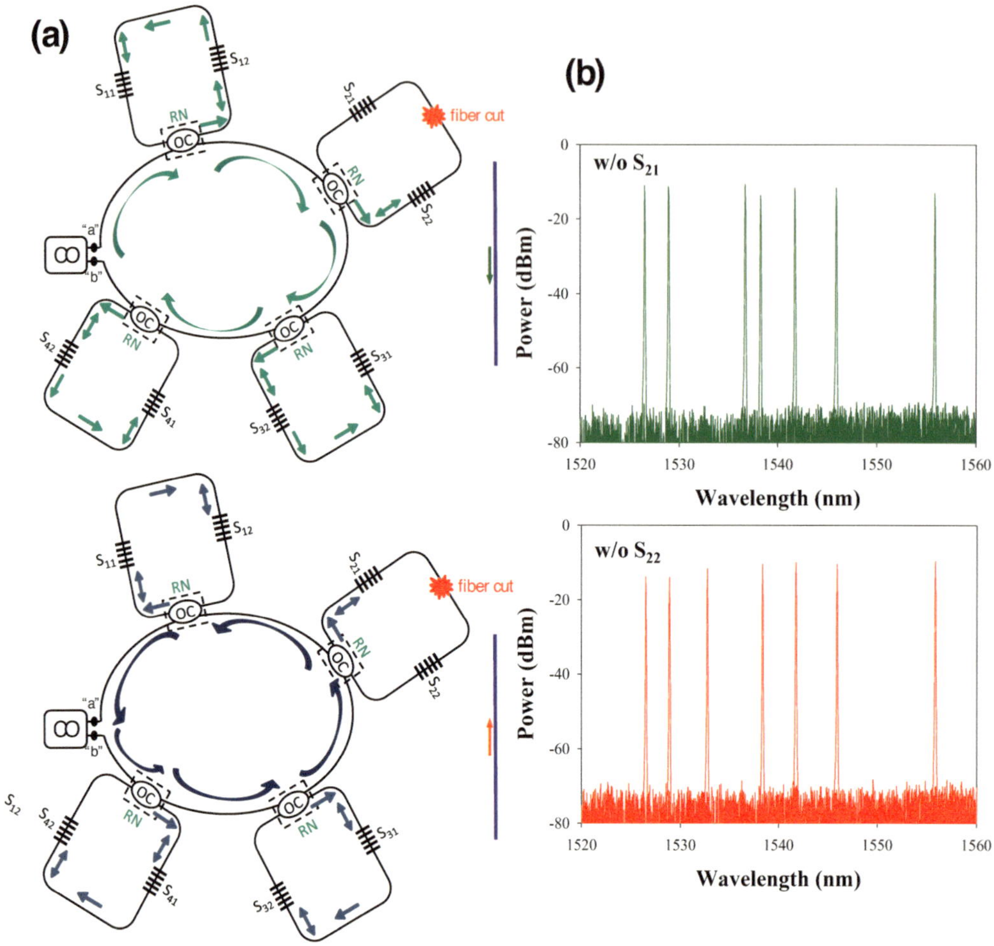

Figure 10. (a) When the sensor system has a fiber cut on sub-ring sensor group "2" in the proposed sensing network. (b) Received output spectra of the eight sensing FBGs in CO. (Green and red lines represent sensing information through the path "a" and "b", respectively.)

and sensor S_{21}, as shown at the top of Figure 11(a). Thus, at the top of Figure 11(b), the output sensing spectrum lacks the sensor S_{22} and sensor S_{21}. To obtain and detect the disappeared sensors, the CO would control the OS to connect the transmission path "b", as illustrated at the bottom of Figure 11(a). At the bottom of Figure 11(b), the output spectrum of sensor information lacks a sensor S_{22}. Then, by comparing the two output sensing spectra of Figure 11(b), we can easily find out that the fault is on sensor S_{22}. Besides, if two sensors of S_{22} and S32 are cut, the measured output spectra would lack the sensor S_{21}, S_{22}, S_{31}, and S_{32} through the path "a". When the sensing path passes through the path "b", the measured output spectrum would lack the sensor S_{22} and S_{32}. As a result, this proposed sensing network can find out two or more sensor cuts.

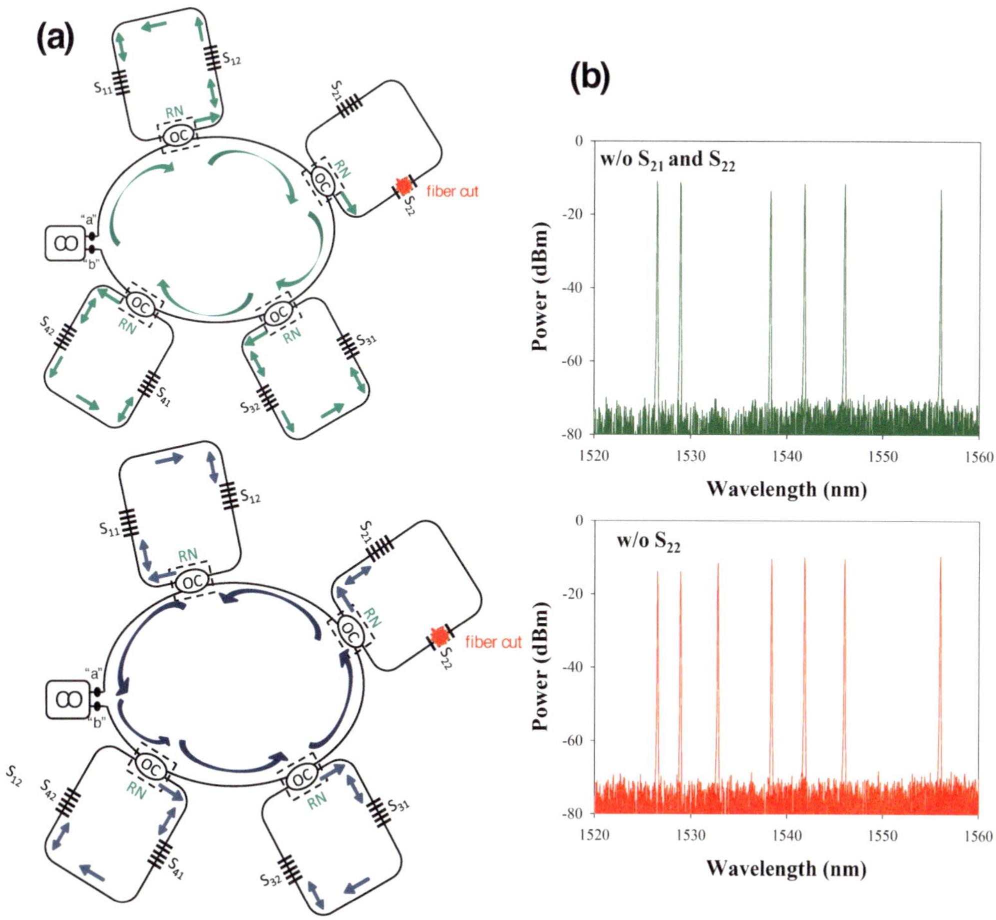

Figure 11. (a) When the sensor system has a fiber cut on the sensor S22 in the proposed sensing network. (b) Received output spectra of the eight sensing FBGs in CO. (Green and red lines represent sensing information through the path "a" and "b", respectively.)

4. Conclusion

In summary, in first proposed design, we have proposed and experimentally investigated a simple self-restored FBG based sensor ring system for long distance sensing. Besides, there is no active component in the FBG sensor architecture for cost reduction. The sensing mechanism is based on a 25 km cavity length EDF ring laser for detecting the multiple FBG sensors in the network. In this experiment, three scenarios of fault detections were experimentally studied, showing that the sensing network survivability, reliability and capacity for the multiple sensors can be enhanced. In the future, this proposed sensing network can be integrated into a fiber access network for advanced applications. And in second proposed design, we have proposed and experimentally investigated a simply self-restored FBG based sensor ring system. There are no active components in the entire sensing architecture. In this experiment, the network survivability and capacity for the multi-point sensor systems are also enhanced. Besides, the TLS is adopted in CO for FBG sensing. The survivability of a eight-point FBG sensor is examined and analyzed. Due to the passive sensor network, the cost-effective and intelligent sensing system is entirely centralized by the CO. As a result, the experimental results show that the proposed system can assist the reliable FBG sensing network for a large-scale and multi-point architecture.

Author details

Chien-Hung Yeh[1] and Chi-Wai Chow[2]

1 Information and Communications Research Laboratories, Industrial Technology Research Institute (ITRI), Chutung, Hsinchu, Taiwan

2 Department of Photonics, National Chiao Tung University, Hsinchu, Taiwan

References

[1] Yeh, C. H., Chow, Chow, C. W., Wang, C. H., Shih, F. Y., Wu, Y. F., and Chi, S. A simple self-restored fiber Bragg grating (FBG)-based passive sensing network. Measurement Science and Technology 2009; 20(4) 043001.

[2] Yeh, C. H., Chow, C. W., Wu, P. C., and Tseng, F. C. A simple fiber Bragg grating-based sensor network architecture with self-protecting and monitoring functions. Sensors 2011; 11(2) 1375–1382.

[3] Peng, P. C., Tseng, H. Y., and Chi, S. (2002): Fiber-Ring Laser-Based Fiber Grating Sensor System Using Self-Healing Ring Architecture. Microwave and Optical Technology Letters 2002; 35(6) 441-444.

[4] Zhang, B., and Kahrizi, M. High-temperature resistance fiber Bragg grating temperature sensor fabrication. IEEE Sensors Journal 2007; 7(4) 86-591.

[5] Zhao, Y., Zhao, H. W., Zhang, X. Y., Meng, Q. Y., and Yuan, B. A Novel double-arched-beam-based fiber Bragg grating sensor for displacement measurement. IEEE Photonics Technology Letters 2008; 20(15) 1296-1298.

[6] Jin, L., Zhang, W., Zhang, H., Liu, B., Zhao, J., Tu, Q., Kai, G., and Dong, X. An embedded FBG sensor for simultaneous measurement of stress and temperature. IEEE Photonics Technology Letters 2006; 18(1) 154-156.

[7] Yeh, C. H., Lin, M. C., Lee, C. C., and Chi, S. Fiber Bragg grating-based multiplexed sensing system employing fiber laser scheme with semiconductor optical amplifier. Japanese Journal of Applied Physics 2005; 44(9) 6590-6592.

[8] Chung, W. H., Tam, H. Y., Wai, P. K. A., and Khandelwal, A. Time- and wavelength-division multiplexing of FBG sensors using a semiconductor optical amplifier in ring cavity configuration. IEEE Photonics Technology Letters 2005; 17(12) 2709-2711.

[9] Yeh, C. H., and Chi, S. Fiber-fault monitoring technique for passive optical networks based on fiber Bragg gratings and semiconductor optical amplifier. Optics Communications 2006; 257(2) 306-310.

[10] Peng, P. C., Lin, W. P., and Chi, S. A self-healing architecture for fiber Bragg sensor network. Proceeding of Sensors 2004; 60-63.

[11] Wu, C. Y., Feng, K. M., Peng, P. C., and Lin, C. Y. Three-dimensional mesh-based multipoint sensing system with self-healing functionality. IEEE Photonics Technology Letters 2010; 22(8) 565-567.

[12] ITU-T, Recommendation G. 984.1 (2003): Gigabit-capable passive optical network (GPON): General Characteristics.

[13] Zhang, L., Liu, Y., Wiliams, J. A. R., and Bennion, I. Enhanced FBG strain sensing multiplexing capacity using combination of intensity and wavelength dual-coding technique. IEEE Photonics Technology Letters 1999; 11(12) 1638-1641.

Chapter 7

Bio and Chemical Sensors Based on Surface Plasmon Resonance in a Plastic Optical Fiber

Nunzio Cennamo and Luigi Zeni

Additional information is available at the end of the chapter

http://dx.doi.org/10.5772/57148

1. Introduction

Surface Plasmon Resonance (SPR) is known to be a very sensitive technique for determining refractive index variations at the interface between a metallic layer and a dielectric medium (analyte). SPR is widely used as a detection principle for many sensors operating in different application fields, such as bio and chemical sensing. In practical implementations, the biological targets are usually transported through a microfluidic system by means of a buffer fluid or a carrier fluid. In SPR sensors, the transducing media (ligands) are usually bonded on the metallic layer surface so that, when they react with the target molecules present in the analyte, the refractive index at the outer interface changes, and this change is detected by suitable optical interrogation. In the scientific literature, many different configurations based on SPR in silica optical fibers, are usually found [1,2].

In general, the optical fiber employed is either a glass one or a plastic one (POF). For low-cost sensing systems, POFs are especially advantageous due to their excellent flexibility, easy manipulation, great numerical aperture, large diameter, and the fact that plastic is able to withstand smaller bend radii than glass. The advantages of using POFs is that the properties of POFs, that have increased their popularity and competitiveness for telecommunications, are exactly those that are important for optical sensors based on optical fibers. Moreover, a further advantage of POF sensors is that they are simpler to manufacture than those made using silica optical fibers. In the scientific literature only simple POF sensors, based on laterally polished bent sections prepared along a plastic optical fiber, can be usually found.

In this chapter, POF sensor configurations are presented in order to monitor an aqueous environment (refractive index around 1.333), with a resolution ranging from 10^{-4} to 10^{-3} (RIU). The classic geometries of sensors based on SPR in silica optical fiber are adapted and borrowed for POF, so representing a simple approach to low cost plasmonic sensing.

The planar gold layer of the sensors, the low resolution and refractive index ranging from 1.332 to 1.420 are three good factors for forthcoming bio/chemical sensors implementation.

Another aspect to be considered is that most often the SPR bio-chemical sensor system is based on a high refractive index prism coated with a thin metallic layer. The incidence angle of the light can be changed in a wide range and, as a consequence, the surface plasma waves (plasmons) may exist whatever the surrounding medium, i.e. a gas or a liquid. These sensors are usually bulky and require expensive optical equipment, not easy to be miniaturized. In addition, the remote sensing can be very difficult to exploit. On the contrary, the use of a POF makes the remote sensing straightforward, and may reduce the cost and dimension of the device, with the possibility of integration of SPR sensing platform with optoelectronic devices, eventually leading to "Lab-on-a-chip".

2. SPR phenomenon

In the optical phenomenon of Surface Plasmon Resonance, a metal-dielectric interface supports a p-polarized electromagnetic wave, namely Surface Plasmon Wave (SPW), which propagates along the interface. When the p-polarized light is incident on this metal-dielectric interface in such a way that the propagation constant (and energy) of resultant evanescent wave is equal to that of the SPW, a strong absorption of light takes place as a result of transfer of energy and the output signal exhibits a sharp dip at a particular wavelength known as the resonance wavelength. The so-called resonance condition is given by the following expression:

$$K_0 n_c \sin\vartheta = K_0 \left(\frac{\varepsilon_{mr} n_s^2}{\varepsilon_{mr} + n_s^2}\right)^{1/2}; \quad K_0 = \frac{2\pi}{\lambda} \tag{1}$$

The term on the left-hand side is the propagation constant K_{inc} of the evanescent wave generated as a result of Attenuated Total Reflection (ATR) of the light incident at an angle θ through a light coupling device (such as prism or optical fiber) of refractive index n_c. The right-hand term is the SPW propagation constant (K_{SP}) with ε_{mr} as the real part of the metal dielectric constant (ε_m) and n_s as the refractive index of the sensing (dielectric) layer. This matching condition of propagation constants is heavily sensitive to even a slight change in the dielectric constant, which makes this technique a powerful tool for sensing of different parameters.

2.1. Spectral mode operation

In SPR sensors with spectral interrogation, the resonance wavelength (λ_{res}) is determined as a function of the refractive index of the sensing layer (n_s). If the refractive index of the sensing layer is altered by δ_{ns}, the resonance wavelength shifts by $\delta\lambda_{res}$. The sensitivity (S_n) of an SPR sensor with spectral interrogation is defined as [3]:

$$S_n = \frac{\delta\lambda_{res}}{\delta n_s}\left[\frac{nm}{RIU}\right] \tag{2}$$

Owing to the fact that the vast majority of the field of an SPW is concentrated in the dielectric, the propagation constant of the SPW is extremely sensitive to changes in the refractive index of the dielectric itself. This property of SPW is the underlying physical principle of affinity SPR bio/chemical sensors. In the case of artificial receptors, as molecular imprinted polymers (MIPs), the polymeric film on the surface of metal selectively recognizes and captures the analyte present in a liquid sample so producing a local increase in the refractive index at the metal surface. The refractive index increase gives rise to an increase in the propagation constant of SPW propagating along the metal surface which can be accurately measured by optical means. The magnitude of the change in the propagation constant of an SPW depends on the refractive index change and its overlap with the SPW field. If the binding occurs within the whole depth of the SPW field, the binding-induced refractive index change produces a change in the real part of the propagation constant, which is directly proportional to the refractive index change.

The resolution (Δn) of the SPR-based optical sensor can be defined as the minimum amount of change in refractive index detectable by the sensor. This parameter (with spectral interrogation) definitely depends on the spectral resolution ($\delta\lambda_{DR}$) of the spectrometer used to measure the resonance wavelength in a sensor scheme. Therefore, if there is a shift of $\delta\lambda_{res}$ in resonance wavelength corresponding to a refractive index change of δn_s, then resolution can be defined as:

$$\Delta n = \frac{\delta n_s}{\delta\lambda_{res}}\delta\lambda_{DR}\left[RIU\right] \tag{3}$$

The Signal-to-Noise Ratio of an SPR sensor depends on how accurately and precisely the sensor can detect the resonance wavelength and hence, the refractive index of the sensing layer. This accuracy in detecting the resonance wavelength further depends on the width of the SPR curve.

The narrower the SPR curve, the higher the detection accuracy. Therefore, if $\delta\lambda_{SW}$ is the spectral width of the SPR response curve corresponding to some reference level of transmitted power, the detection accuracy of the sensor can be assumed to be inversely proportional to $\delta\lambda_{SW}$.

The signal-to-noise ratio of the SPR sensor with spectral interrogation is, thus, defined as:

$$SNR(n) = \left(\frac{\delta\lambda_{res}}{\delta\lambda_{SW}}\right)_n \tag{4}$$

where $\delta\lambda_{SW}$ can be calculated as the full width at half maximum of the SPR curve (FWHM). SNR is a dimensionless parameter strongly dependent on the refractive index changes.

2.2. Amplitude mode operation

In SPR sensors, it can be used a simpler scheme with a monochromatic light source and an optical power meter in the amplitude mode operation, because in the shoulders of spectral response curve, appreciable intensity differences are present, as the given refractive index changes, due to the shift of the curve itself.

In amplitude mode operation the sensitivity (S_n) of an SPR sensor is defined as:

$$S_n = \frac{\delta I_{norm}}{\delta n_s}\left[\frac{a.u.}{RIU}\right] \tag{5}$$

where I_{norm} is the relative output, normalized to a reference level, in order to compensate for light source fluctuations.

The resolution (Δn) of the SPR-based optical sensor, in amplitude mode operation, can be defined as:

$$\Delta n = \frac{\delta n_s}{\delta I_{norm}} \sigma\left[RIU\right] \tag{6}$$

where σ is the standard deviation of the relative output.

3. A POF sensor system for biosensor implementation

The fabricated optical sensor system was realized removing the cladding of a plastic optical fiber along half the circumference, spin coating on the exposed core a buffer of Microposit S1813 photoresist, and finally sputtering a thin gold film using a sputtering machine [4].

The plastic optical fiber has a PMMA core of 980 μm and a fluorinated polymer cladding of 20 μm. The experimental results indicate that the configuration with a fiber diameter of 1000 μm exhibits better performance in terms of sensitivity and resolution but not in terms of SNR [5], as shown in section 5.2.

The refractive index, in the visible range of interest, is about 1.49 for PMMA, 1.41 for fluorinated polymer and 1.61 for Microposit S1813 photoresist. The sample consisted in a plastic optical fiber without jacket embedded in a resin block, with the purpose of easing the polishing process. The polishing process was carried out with a 5 μm polishing paper in order to remove the cladding and part of the core. After 20 complete strokes following a "8-shaped" pattern (according to the manufacturer recommendations, as shown in Figure 1) in order to completely expose the core, a 1 μm polishing paper was used for another 20 complete strokes following a "8-shaped" pattern. The realized sensing region was about 10 mm in length.

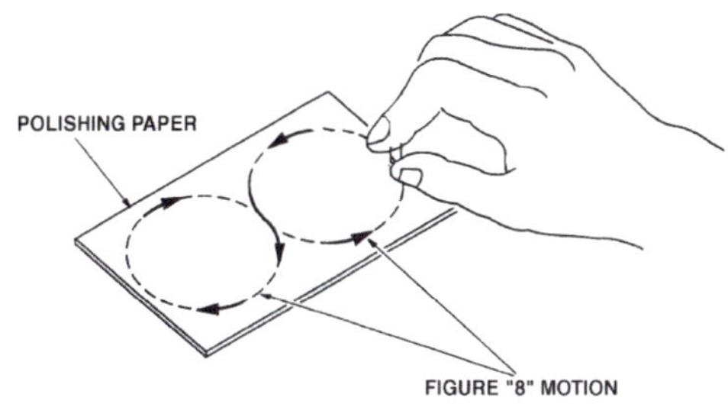

Figure 1. "8-shaped" pattern for POF polishing.

The buffer of Microposit S1813 photoresist was realized by means of spin coating. The Microposit S1813 photoresist is deposited in one drop (about 0.1 ml) on the center of the substrate. The sample is then spun at 6,000 rpm for 60 seconds. The final thickness of photoresist buffer was about 1.5 μm.

As it will be shown in the section 5.1, the experimental results indicate that this configuration with the photoresist buffer layer exhibits better performance in terms of detectable refractive index range and SNR [4].

Finally, a thin gold film was sputtered by using a sputtering machine (Bal-Tec SCD 500).The sputtering process was repeated twice with a current of 60 mA for a time of 35 seconds (20 nm for step). The gold film so obtained was 60 nm thick and presented a good adhesion to the substrate, verified by its resistance to rinsing in de-ionized water.

This sensor based on SPR in a POF is a common tool for surface interaction analysis and biosensing, widely used as a detection principle for sensors that operate in different areas of bio and chemical sensing as reported in several recent review papers [6,7]. In this case on the gold surface there is a bio or chemical layer for the selective detection and analysis of analyte in aqueous solution (see figure 2).

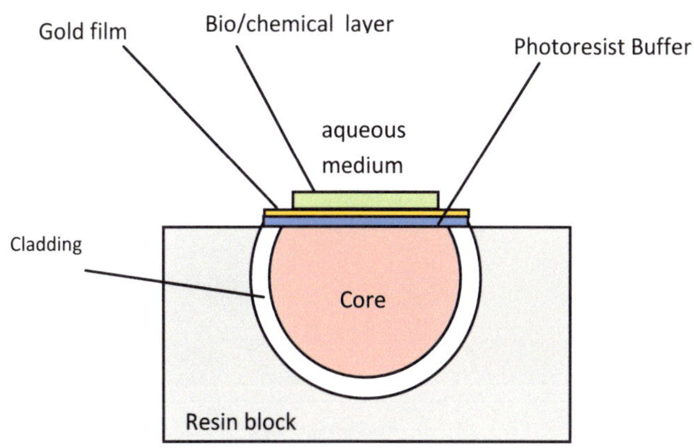

Figure 2. Section of POF sensor

4. Experimental configurations for SPR sensors in plastic optical fibers

The experimental measurements for the characterization of the POF sensor, presented in the previous section, were carried out in two different ways: spectral and amplitude mode. Figure 3 shows the experimental setup arranged to measure the transmitted light spectrum and was characterized by a halogen lamp, illuminating the optical sensor system, and a spectrum analyzer. The employed halogen lamp exhibits a wavelength emission range from 360 nm to 1700 nm, while the spectrum analyzer detection range was from 200 nm to 850 nm. The spectral resolution of the spectrometer was 1.5 nm (FWHM).

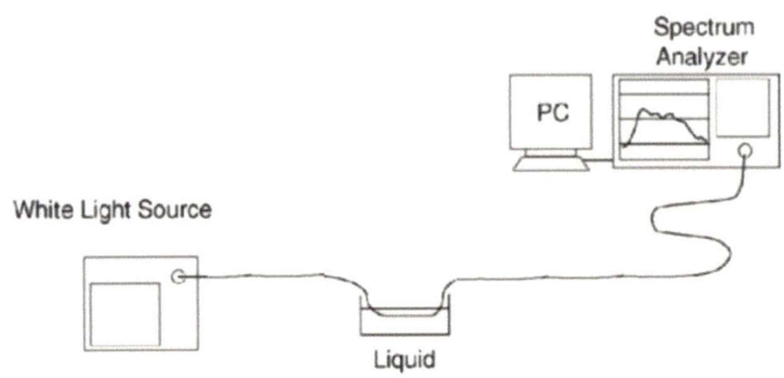

Figure 3. Setup to measure the transmitted light spectrum

Figure 4 shows the measurements carried out, obtaining SPR transmission spectra, normalized to the spectrum achieved with air as the surrounding medium, for three different water-glycerin solutions with refractive index 1.333, 1.351, 1.371, respectively. The sensitivity is calculated as the slope of the resonance wavelength versus refractive index curve, for three refractive index values. In figure 5 the experimental data and the linear fitting are presented.

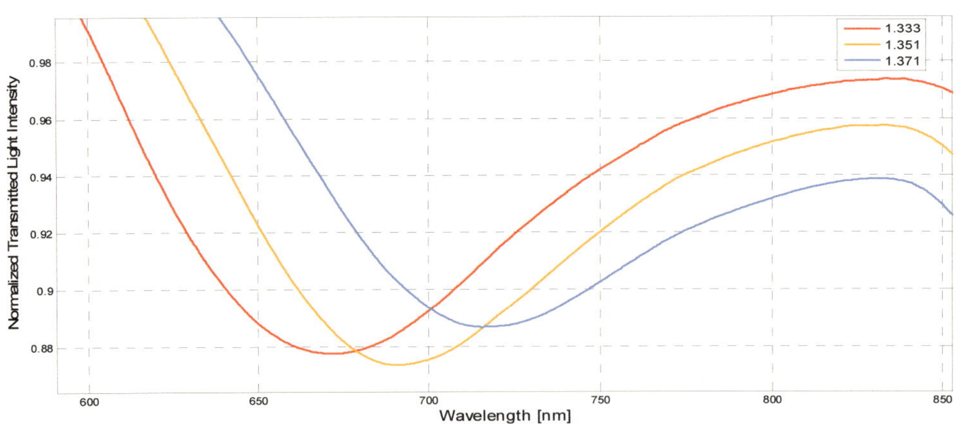

Figure 4. Normalized transmitted light intensity as function a function of the wavelength

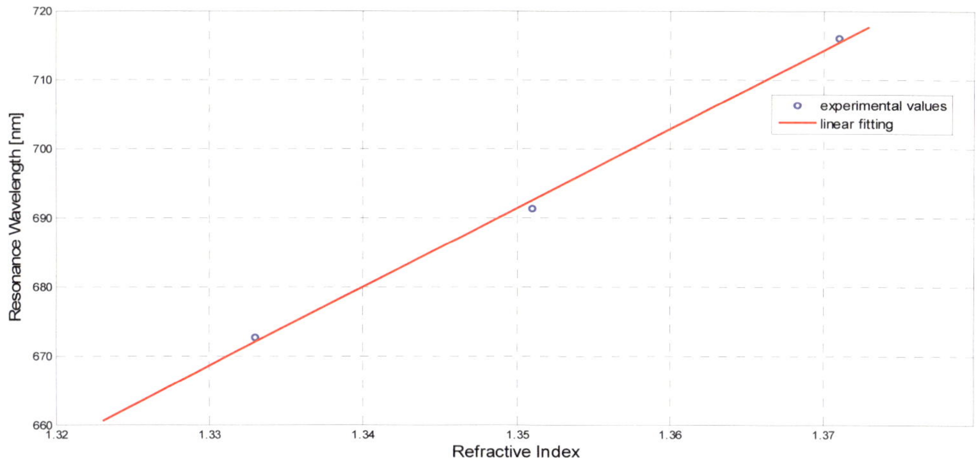

Figure 5. Resonance wavelength as a of the refractive index

Figure 6 shows the second setup. It is composed of an LED, whose wavelength is 670nm, as the light source, a beam splitter, two photodiodes whose function is to convert the light into an electrical signal and an oscilloscope for signal acquisition connected to a PC.

In figure 7, the relative output is plotted as a function of time for three different refractive index values.

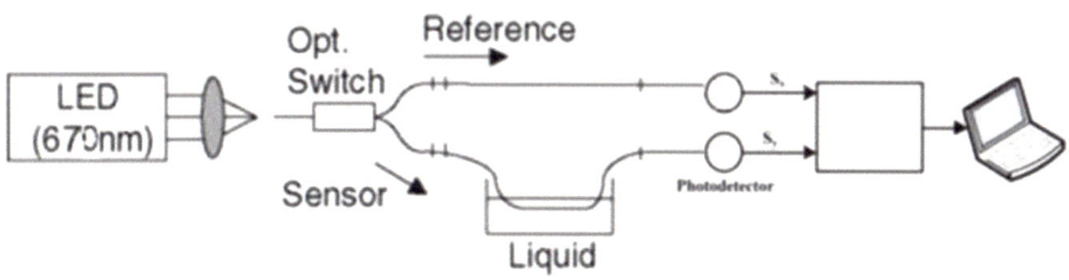

Figure 6. Setup using a LED and two photodiode

According to an experiment for the evaluation of system stability, the standard deviation of the relative output was found to be 3.84×10^{-4} for the 8 min continuous operation in the air with the intention of circumventing liquid evaporation. In figure 8, the relative output for three different refractive index and the linear fitting to the experimental values are presented.

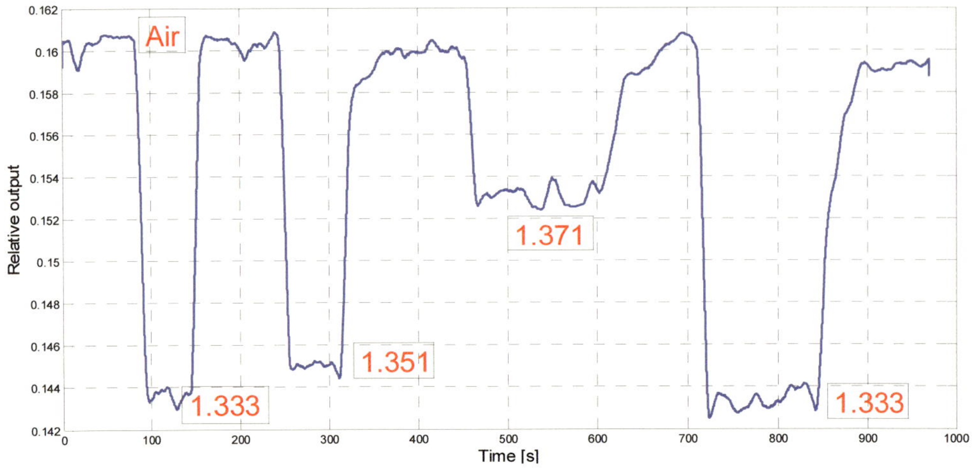

Figure 7. Relative output as a function of time, for different refractive index values

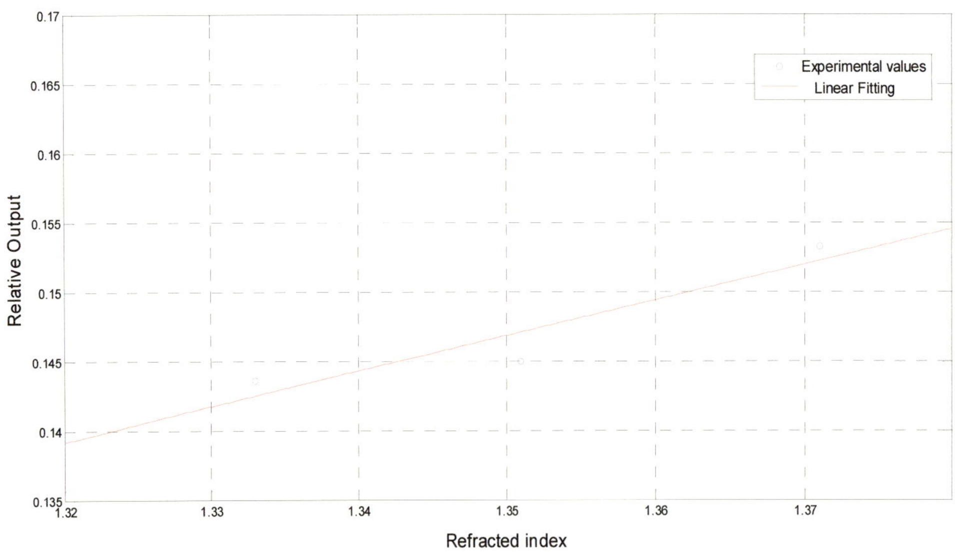

Figure 8. Relative output as a function of refractive index

For a clearer comparative analysis of the two experimental setups, table 1 summarizes the average values of the experimentally measured performance parameters, evaluated by Matlab software, for external medium refractive index ranging from 1.333 to 1.371.

Experimental setup	Sensitivity (S)	Margin of error (∂E)	$\Delta n = S^{-1} * \partial E$
White Light Source/ spectrum analyzer	$\frac{\delta\ resonance\ wavelength}{\delta\ refractive\ index} = 1.14 * 10 \wedge 3 \left[\frac{nm}{RIU}\right]$	(spectral resolution of the spectrometer) 1.5 nm (FWHM)	0.00131 [RIU]
LED/photodiodes	$\frac{\delta\ \text{relative output}}{\delta\ \text{refractive index}} = 0.26 \left[\frac{\text{a.u.}}{\text{RIU}}\right]$	(standard deviation) 3.84×10^{-4}	0.00147 [RIU]

Table 1. Performance parameters for different experimental configurations

5. Performance comparison of sensors based on SPR in POF

In this section different POF sensor configurations are presented and experimentally tested with spectral interrogation: First, the configurations with and without photoresist buffer layer; then, the configurations with two different POF core diameters and finally the configuration with a tapered POF.

5.1. POF sensors with and without the photoresist layer

In this section we present two configurations with and without the photoresist buffer layer (see figure 2). In the series of performed experiments, water-glycerin solutions were used to achieve an aqueous medium with variable refractive index. Without the buffer layer, in the same operating conditions, the sensor is capable of monitoring refractive indexes ranging from 1.330 to 1.360. In Figure 9 are presented the experimentally obtained SPR transmission spectra, normalized to the spectrum achieved with air as the surrounding medium, for three different water-glycerin solutions with refractive index ranging from 1.332 to 1.352.

In the presence of the photoresist buffer layer, the refractive index range is increased. In particular, this fiber optic sensor is capable of monitoring an aqueous environment whose refractive index ranges from 1.332 to 1.418.

In Figure 10 are presented the experimentally obtained SPR transmission spectra, normalized to the spectrum achieved with air as the surrounding medium, obtained in this case with the photoresist buffer layer, for different water-glycerin solutions with refractive index ranging from 1.332 to 1.418.

In presence of the photoresist buffer layer, the refractive index range is increased while the sensitivity is the same.

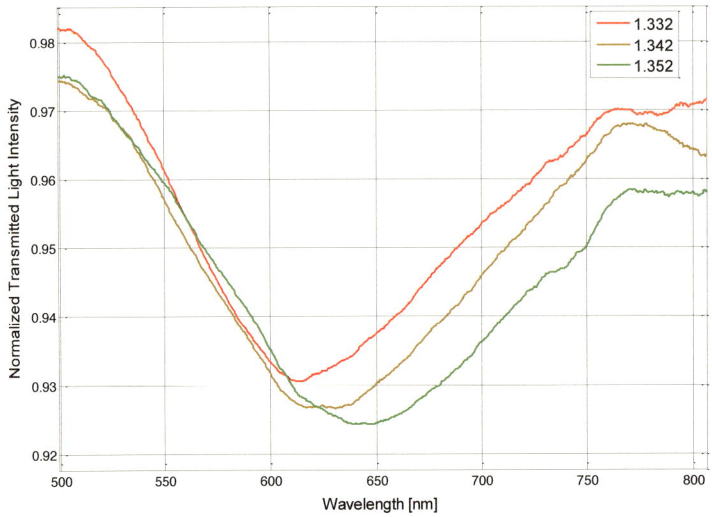

Figure 9. Experimentally obtained SPR transmission spectra, normalized to the air spectrum, for different refractive index of the aqueous medium. Configuration without the photoresist buffer layer.

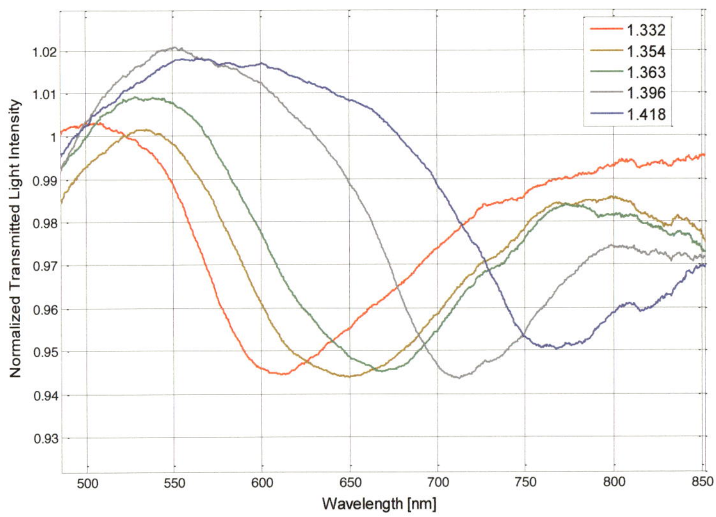

Figure 10. Experimentally obtained SPR transmission spectra, normalized to the air spectrum, for different refractive index of the aqueous medium. Configuration with the photoresist buffer layer.

Without the photoresist buffer layer there is a decrease in the power transmitted to the fiber end facet, due to a greater dissipation. This decrease results in the decrease of the SPR curve and the increase of the SPR curve width. Therefore, it can be conveniently established that SPR curve width increases ($\delta\lambda_{SW}$) without the photoresist buffer layer, as shown for example in Figure 11 for a refractive index equal to 1.332 (water).

http://dx.doi.org/10.5772/57148

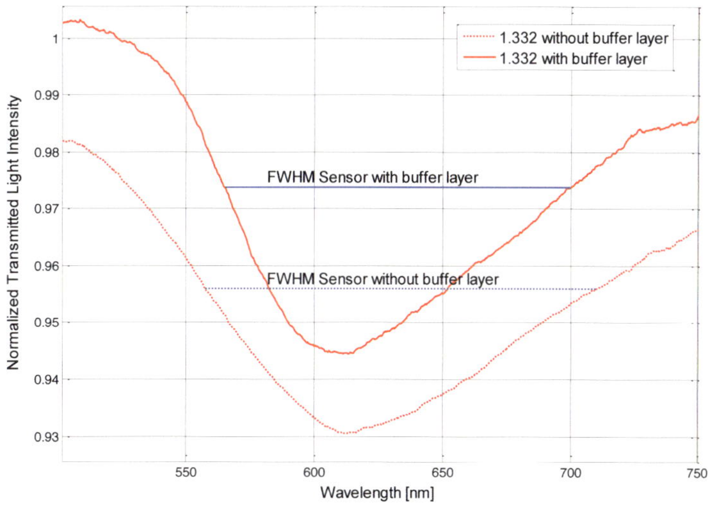

Figure 11. The full width at half maximum (FWHM) of the SPR curve for the two sensors configurations with and without the buffer layer for an external refractive index of 1.332.

The observed absorption band is the result of the convolution of different resonance peaks. Each peak is obtained for a specific resonance condition defined by a given angle-wavelength couple. Therefore, the experimental results indicate that the configuration with the photoresist buffer layer exhibits better performance in terms of detectable refractive index range and SNR.

5.2. POF sensors with two different POF core diameters

In this section we presented the influence of POF core diameter on sensor performances. Before entering the details of the discussion, as a similar analysis is present in the literature with reference to a sensor based SPR in silica optical fiber without any buffer layer between the fiber core and the gold film [8], it is convenient to briefly recall some fundamental aspects of light rays propagation in optical fibers where surface plasmons are excited.

Inside an optical fiber, any light ray making an angle θ from the normal to core-cladding interface undergoes multiple reflections (N_{ref}), depending on the length of SPR sensing region (L) and fiber core diameter (D), according to the following relation [8]:

$$N_{ref}(\theta) = \frac{L}{D \tan \theta} \quad (7)$$

To determine the effective transmitted power, the reflectance (R_e) for a single reflection is raised to the power equal to corresponding number of reflections. Therefore, the generalized expression (all guided rays) for the normalized transmitted power (P_{trans}) in sensors based on SPR in fiber optic can be written as:

$$P_{trans} = \frac{\int_{\theta cr}^{\pi/2} R_e^{N_{ref}(\theta)} I(\theta) d\theta}{\int_{\theta cr}^{\pi/2} I(\theta) d\theta} \tag{8}$$

In Equation (8), I(θ) is the angular intensity distribution corresponding to the light source used. Further, θ_{cr} is the critical angle of the fiber, which heavily depends on the Numerical Aperture (NA) of the fiber and light wavelength.

The angular range from θ_{cr} to π/2 covers whole range of guided rays (or modes) as these angles correspond to the highest order mode and the fundamental mode of an optical fiber, respectively. The number of modes that can propagate in a fiber depends on the fiber's Numerical Aperture (or acceptance angle) as well as on its core diameter and the wavelength of the light. For a step-index multimode fiber, the number of such modes, M, is approximated (M >> 1) by:

$$M \cong 0.5 * \left(\frac{\pi * D * NA}{\lambda} \right)^2 \tag{9}$$

where D is the core diameter, λ is the operating wavelength, NA is the Numerical Aperture (or acceptance angle, as shown in Figure 12).

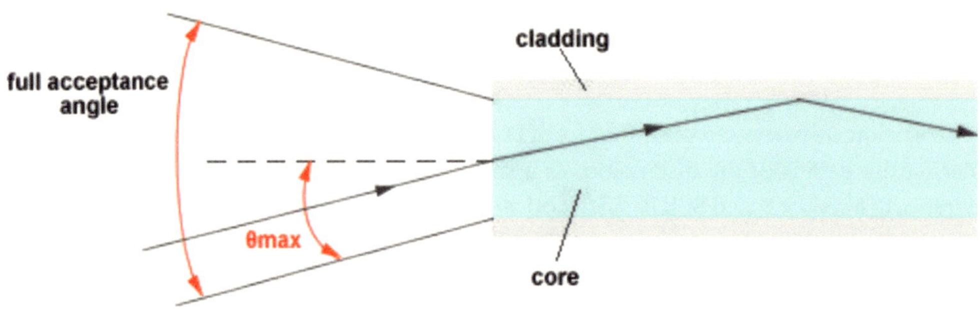

Figure 12. Optical fiber acceptance cone.

In general, numerical aperture of a Plastic Optical Fiber is greater than that of a Silica Optical Fiber. The resonance condition (see Equation (1)) is satisfied at different wavelengths depending on which combination of core diameter and sensing region length is considered. It is so clear that the performance parameters of a fiber optic SPR sensor strictly depend on the values of design parameters such as fiber core diameter (D), sensing region length (L), and numerical aperture (NA). In this section, we analyze the influence of two values of Plastic Optical Fiber core diameter (D) on the performance of a sensor based on Surface Plasmon Resonance in a POF, where the sensing region length is fixed and a photoresist buffer layer is placed between the fiber core and the gold film (see figure 2).

http://dx.doi.org/10.5772/57148

For sensors based on SPR in optical fiber (silica or plastic) the shift in resonance wavelength ($\delta\lambda_{res}$), for a fixed refractive index variation (δn_s), increases with a decrease in the number of reflections. Therefore, sensitivity increases with the increase of fiber core diameter and with the decrease of sensing region length.

Figure 13 reports the experimentally obtained SPR transmission spectra, when the diameter POF is 1,000 μm (figure 13 (a)) and 250 μm (figure 13 (b)), normalized to the spectrum achieved with air as the surrounding medium, for five different water-glycerin solutions with refractive index ranging from 1.332 to 1.372.

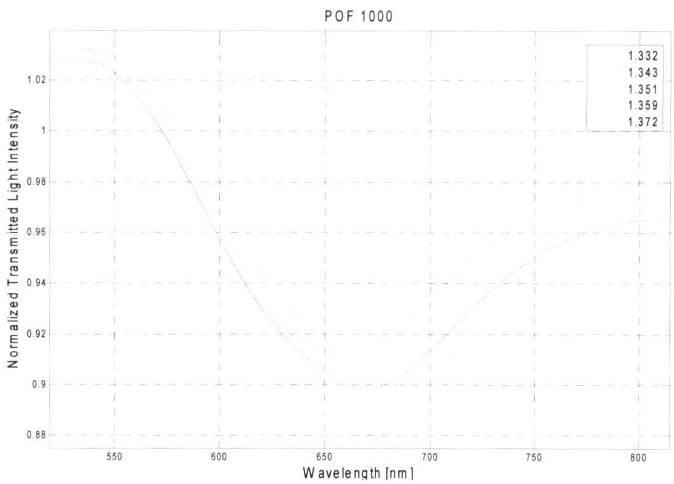

(a)

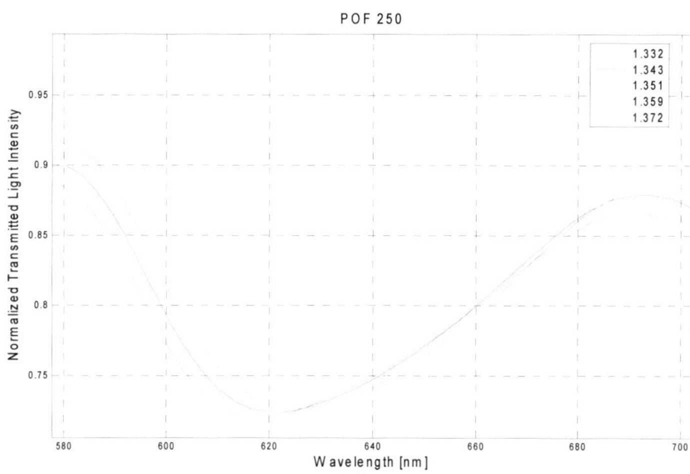

(b)

Figure 13. Experimentally obtained SPR transmission spectra, normalized to the air spectrum, for different refractive index of the aqueous medium. (a) Configuration with a 1,000 μm diameter POF; (b) Configuration with a 250 μm diameter POF

Figure 14 shows the resonance wavelength versus the refractive index obtained with the two different configurations. In the same figure is also presented the linear fitting to the experimental data, showing a good linearity for the sensors. The Pearson's correlation coefficient (R) is 0.99 for the sensor with a POF of 1,000 μm and 0.98 for the sensor with a POF of 250 μm diameter.

The sensitivity, as defined in Equation (2), is the shift of the resonance wavelength (nm) per unit change in refractive index (nm/RIU). Therefore, it is the angular coefficient of the linear fitting. Figure 14 shows as the sensitivity increases with the fiber core diameter.

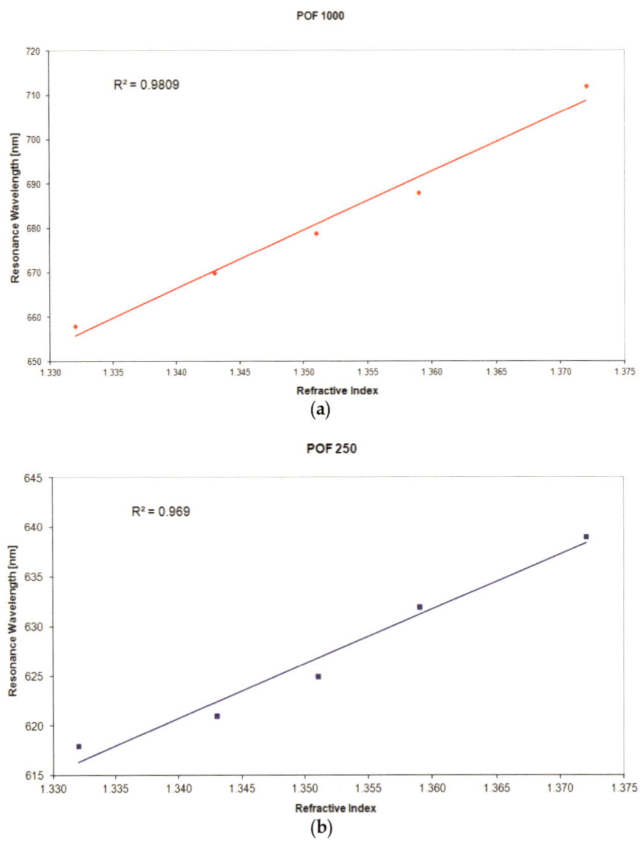

Figure 14. Plasmon resonance wavelength as a function of the refractive index. (a) Configuration with a 1,000 μm diameter POF. (b) Configuration with a 250 μm diameter POF.

Furthermore, as sensor's resolution also depends on the variation of $\delta\lambda_{res}$ (see Equation (3)), therefore, similarly to sensitivity, resolution also tends to improve for larger fiber core diameters (see Figure 15). In fact, the resolution (Δn) can be calculated as the angular coefficient of the linear fitting in Figure 15 multiplied to the spectral resolution ($\delta\lambda_{DR}$) of the spectrometer used to measure the resonance wavelength.

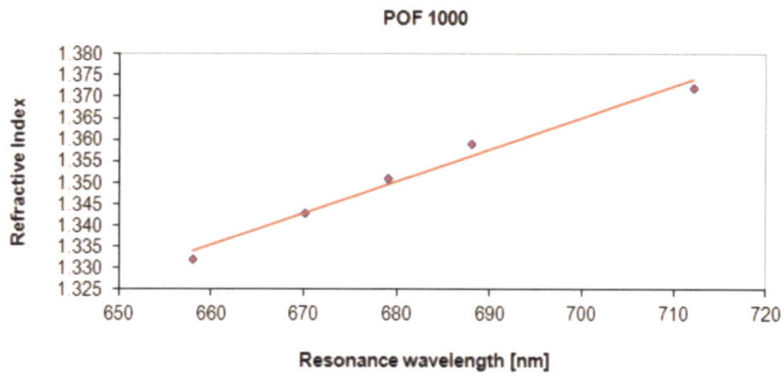

(a)

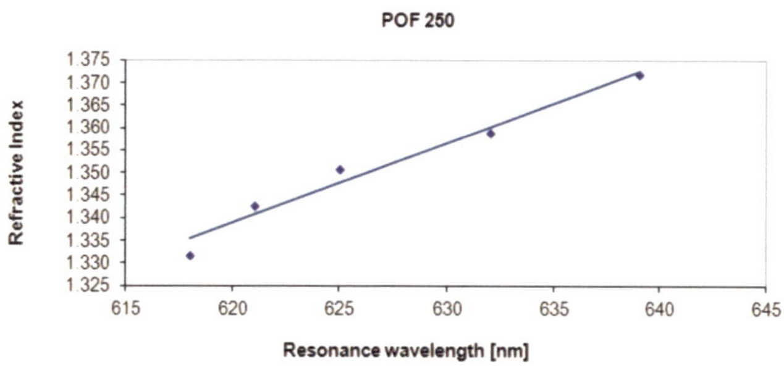

(b)

Figure 15. Refractive index as a function of the plasmon resonance wavelength. (a) Configurations with a 1,000 μm diameter POF. (b) Configurations with a 250 μm diameter POF.

The experimental results obtained with the two values of POF core diameter have shown as the Numerical Aperture of POF and the photoresist buffer layer have produced a different behavior with respect to many different configurations based on SPR in silica optical fibers, as SNR is concerned. From our experimental results [5], it is clear that the shift in resonance wavelength ($\delta\lambda_{res}$), for a fixed refractive index variation (δ_{ns}), increases when the core diameter increases. Therefore, sensitivity increases with an increase in fiber core diameter. Furthermore, in the sensors based on SPR in POF (configuration with the photoresist buffer layer) as already established, SPR curve width ($\delta\lambda_{SW}$) increases with an increase in fiber core diameter. Therefore, it can be conveniently established that SPR curve width increases ($\delta\lambda_{SW}$) with the increase of fiber core diameter, as shown in Figure 16 for a refractive index equal to 1.332.

SPR curve width $\delta\lambda_{SW}$ can be calculated as the full width at half maximum (FWHM) of the SPR curve. FWHM of the SPR curve as a function of the refractive index is shown in Figure 17. Therefore, SNR is expected to decrease because an increase in the shift in resonance wavelength ($\delta\lambda_{res}$) produces a larger increase in the curve width ($\delta\lambda_{SW}$), for a fixed increase in fiber core diameter.

More precisely, for a POF with 250 μm of diameter, the angular coefficient of the linear fitting shown in Figure 14 ($\delta\lambda_{res}$) is greater than the angular coefficient of the linear fitting shown in Figure 17 ($\delta\lambda_{SW}$). In this case SNR is greater than one. For a POF with 1,000 μm of diameter the angular coefficient of the linear fitting shown in Figure 14 ($\delta\lambda_{res}$) is lower than the angular coefficient of the linear fitting shown in Figure 17 ($\delta\lambda_{SW}$). In this case SNR is less than one.

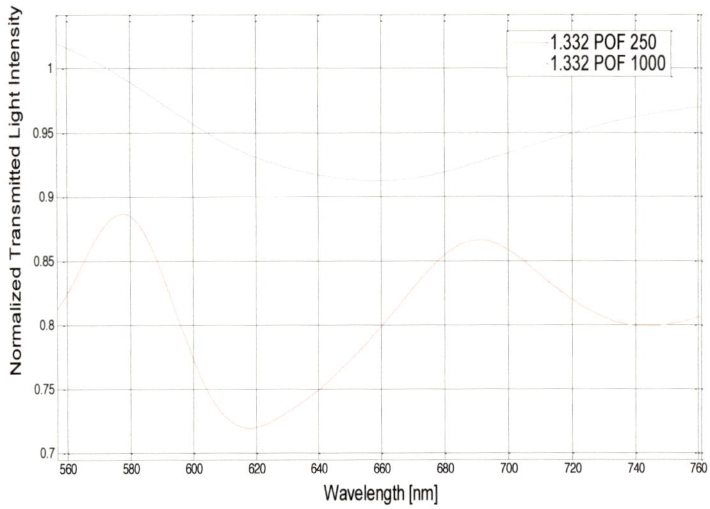

Figure 16. The full width at half maximum of the SPR curve for the two sensors configurations (250 μm and 1,000 μm POF diameter) for an external refractive index of 1.332.

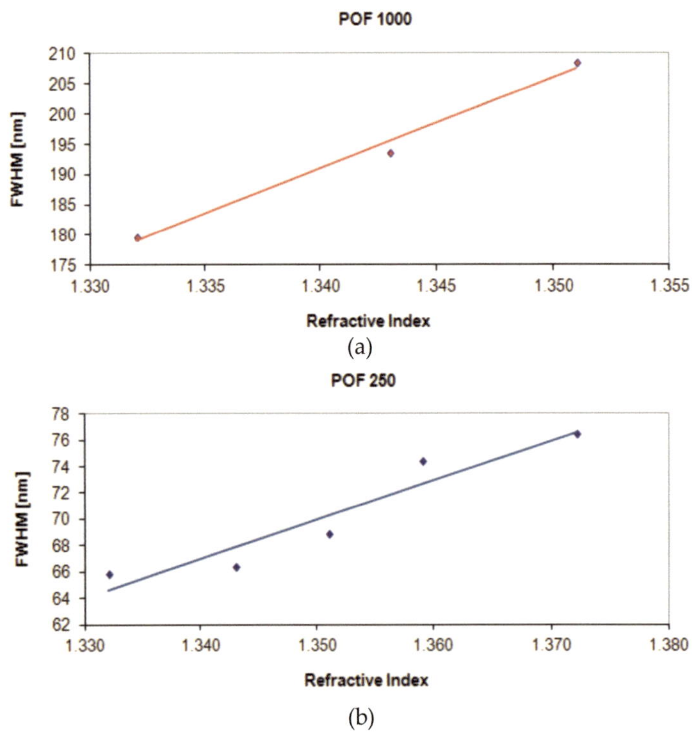

Figure 17. The full width at half maximum of the SPR curve as a function of the refractive index. (a) Configuration with a 1,000 μm diameter POF. (b) Configuration with a 250 μm diameter POF.

The plasmon resonance wavelength as a function of the full width at half maximum of the SPR curve is shown in Figure 18. SNR can be calculated as the angular coefficient of the linear fitting reported in Figure 18. From the above figure, it is clear that the SNR increases when the fiber core diameter decreases. It is important to emphasize that the calculation, from experimental data, of the single values of above parameters has been carried out by employing a first-order approach, while the linear fitting does not imply an actual linear relationship but it is just a way to extrapolate a trend and allow an easy comparison between the two sensor systems.

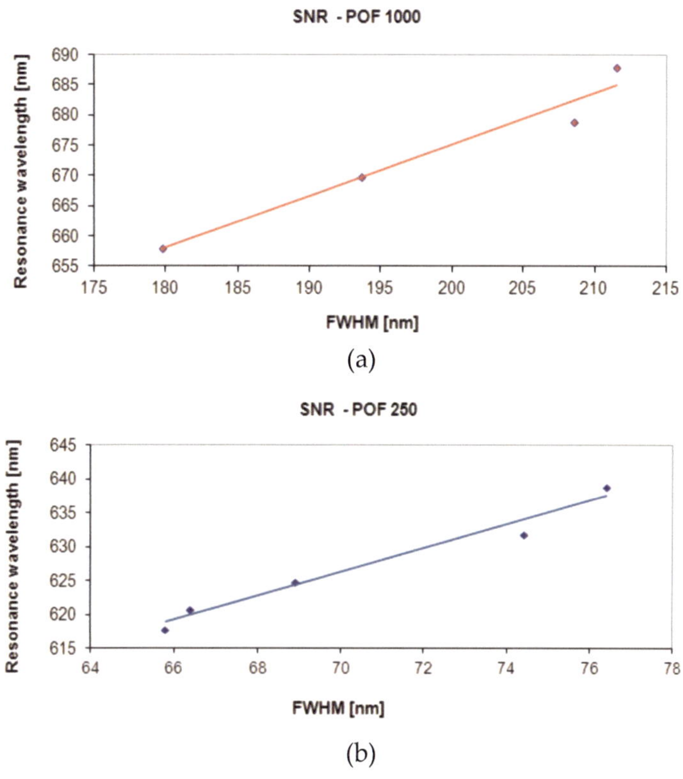

Figure 18. Plasmon resonance wavelength as a function of the full width at half maximum of the SPR curve. (a) Configuration with a 1,000 μm diameter POF. (b) Configuration with a 250 μm diameter POF.

For a clearer comparative analysis between the two sensors with 250 μm and 1,000 μm diameter POFs, Table 2 summarizes the averages values of the experimentally measured performance parameters, evaluated by Matlab software, for external medium refractive index ranging from 1.332 to 1.372.

POF Diameter [μm]	Resolution (Δn) [RIU]	Signal-to-noise ratio (SNR)	Sensitivity (S) [nm/RIU]	FWHM/Δn [nm/RIU]
250	0.0027	1.7548	0.549×10^3	0.298×10^3
1,000	0.0010	0.8569	1.325×10^3	1.495×10^3

Table 2. Performance comparison for the two sensors configurations: 250 μm and 1,000 μm diameter POF, respectively.

5.3. Sensor configuration with taperd POF

Figure 19 shows the optical sensor configuration with a tapered POF. The optical sensor can be realized removing the cladding of a plastic optical fiber along half circumference, heating and stretching it and finally sputtering a thin gold film.

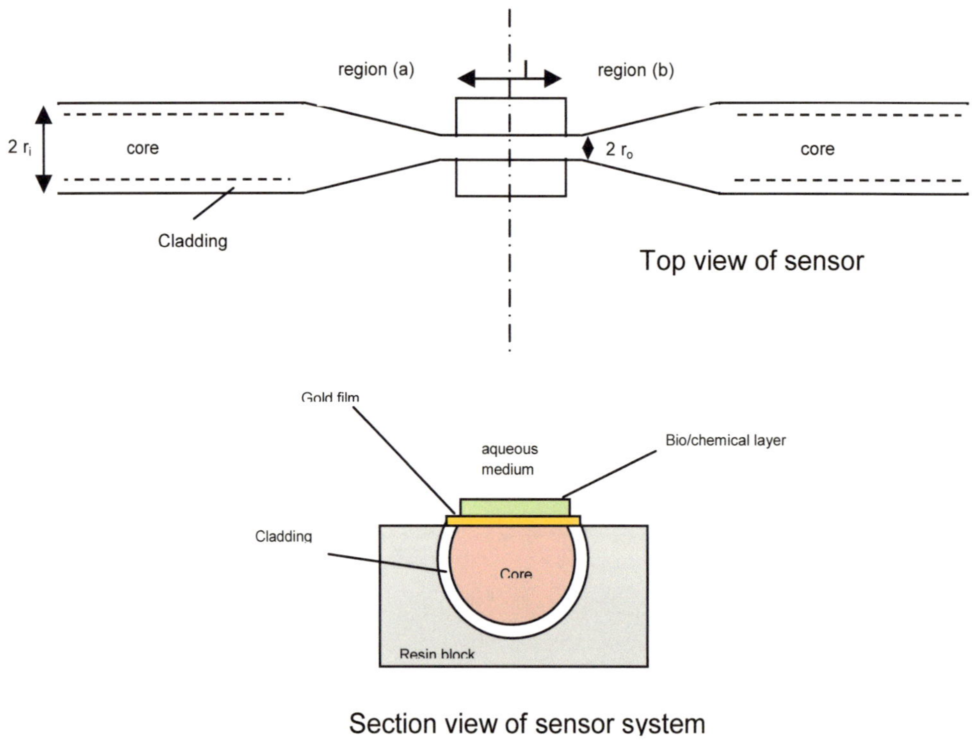

Figure 19. Sensor system based on SPR in tapered POF

The experimental results, presented in this section, are obtained with the following configuration: The plastic optical fiber has a PMMA core of 980 μm and a fluorinated cladding of 20μm. The taper ratio (r_i/r_o) is about 1.5 and the sensing region (L) is about 10 mm in length. The thicknesses of gold layer is about 60 nm. The sensor was realized starting from a plastic optical fiber, without protective jacket, heated (at 150°C) and stretched with a motorized linear positioning stage until the taper ratio reached 1.5. After this step, the POF was embedded in a resin block, and polished with a 5 μm polishing paper in order to remove the cladding and part of the core. After 20 complete strokes following a "8-shaped" pattern in order to completely expose the core, a 1 μm polishing paper was used for another 20 complete strokes with a "8-shaped" pattern. The thin gold film was sputtered by using a sputtering machine (Bal-Tec SCD 500).The sputtering process was repeated three-time with a current of 60 mA for a time of 35 seconds (20 nm for step). On the top of planar gold film it is possible to apply a bio/chemical layer for the selective detection of analytes.

When SPR is achieved in optical fibers, the introduction of the tapered region (a) in fig. 19 helps to reduce the incidence angles of the guided rays in the fiber close to the critical angle of the unclad uniform tapered region. This is obtained by choosing the minimum allowed value of the radius of the output end of the taper so that all the rays remain guided in the uniform core sensing region [9]. After propagating through the uniform region the rays enter the tapered region (b) (see fig. 19) which reconverts the angles of these rays into their initial values so that they can propagate up to the output end of the fiber. Thus, the sensing probe has the minimum diameter such that no ray leaks out and a majority of rays are bound to propagate close to the critical angle, thereby increasing the penetration depth of the evanescent field to almost the maximum value [9].

Figure 20 reports the experimentally obtained SPR transmission spectra, obtained with this tapered POF configuration, normalized to the spectrum achieved with air as the surrounding medium, for six different water-glycerin solutions with refractive index ranging from 1.333 to 1.385.

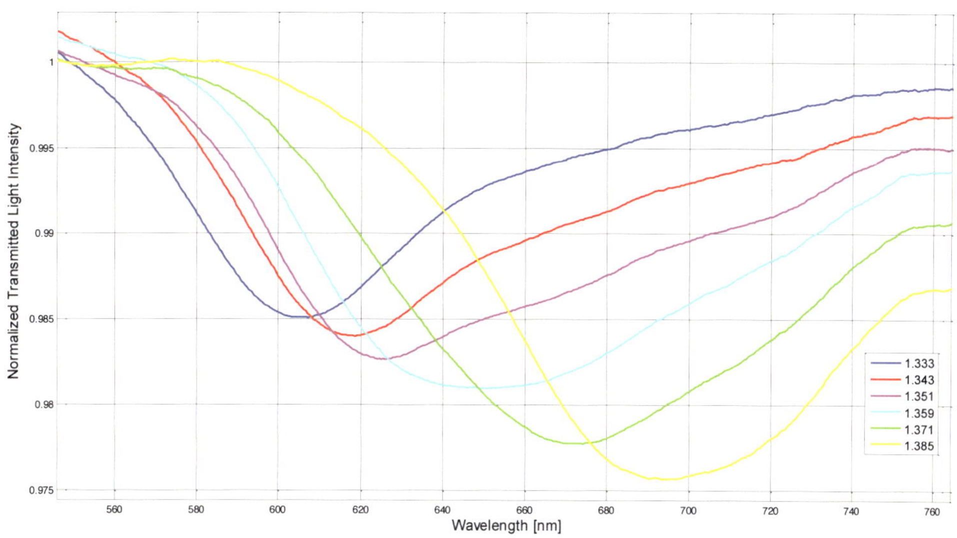

Figure 20. Experimentally obtained SPR transmission spectra, normalized to the air spectrum, for different refractive index of the aqueous medium. Configuration with tapered POF.

Figure 21 shows the resonance wavelength versus the refractive index. In the same figure, it is also presented the linear fitting to the experimental data. The sensitivity, as defined in Equation (2), is the angular coefficient of the linear fitting. Figure 21 shows as the sensitivity increases with the tapered POF configuration.

In this case, i.e. tapered POF without photoresist buffer layer, the sensitivity is about $2*10^3$ (nm/RIU), and it is doubled with respect to the case without tapered POF configuration.

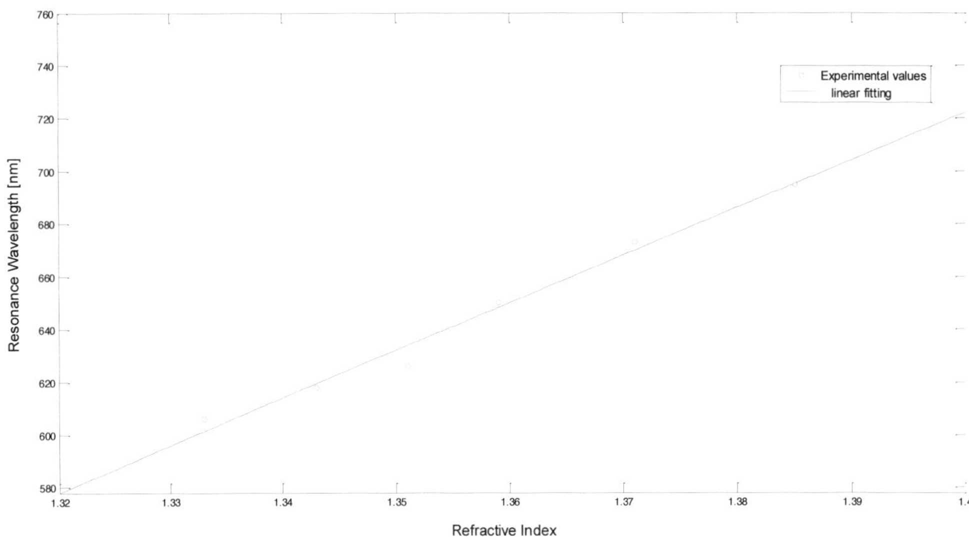

Figure 21. Plasmon resonance wavelength as a function of the refractive index. Configuration with tapered POF

6. SPR for detection of Bio/chemical Analytes

When artificial receptors are used for Bio/chemicals detection, the film on the surface of metal selectively recognizes and captures the analyte present in a liquid sample so producing a local increase in the refractive index at the metal surface.

The refractive index change Δn_s induced by the analyte molecules binding to the biorecognition elements can be expressed as [10]:

$$\Delta n_s = \left(\frac{dn}{dc}\right)_{vol} \Delta c_s \tag{10}$$

where $(dn/dc)_{vol}$ is the volume refractive index increment, and Δc_s is the concentration variation of bound analyte expressed in mass/volume or in any other concentration units in the polymer phase. The value of the refractive index increment depends on the structure of the analyte molecules [11,12].

The refractive index increase gives rise to an increase in the propagation constant of SPW propagating along the metal surface which can be accurately measured, as previously stated.

For this bio-chemical optical sensor with spectral interrogation, the sensitivity is more conveniently defined as:

$$S = \frac{\delta\lambda_{res}}{\delta C}\left[\frac{nm}{M}\right] \tag{11}$$

In other words, the sensitivity can be defined by calculating the shift in resonance wavelength per unit change in analyte concentration (nm/M).

7. Conclusions

In this chapter we have presented an analysis of SPR phenomenon, a POF sensor based on SPR with the related experimental configurations, a performance comparison of sensors based on SPR in POF and a possible implementation as biosensors. The presented devices are based on the excitation of surface plasmons at the interface between an under test medium (aqueous medium) and a thin planar gold layer deposited on a modified plastic optical fiber. Therefore, the proposed sensing head, being low cost and relatively easy to realize, may be very attractive for bio/chemical sensor implementation [6,7].

Author details

Nunzio Cennamo and Luigi Zeni

Department of Industrial and Information Engineering, Second University of Naples, Aversa, Italy

References

[1] J. Homola, Present and future of surface plasmon resonance biosensors, Anal. Bioanal. Chem. 377, (2003) 528–539.

[2] R.C. Jorgenson, S.S. Yee, A fiber-optic chemical sensor based on surface plasmon resonance, Sens. Actuators B: Chem. 12, (1993) 213–220.

[3] M. Kanso, S. Cuenot, G. Louarn, Sensitivity of optical fiber sensor based on surface plasmon resonance: Modeling and experiments, Plasmonics 3, (2008) 49–57.

[4] N. Cennamo, D. Massarotti, L. Conte, L. Zeni, Low cost sensors based on SPR in a plastic optical fiber for biosensor implementation, Sensors 11, (2011) 11752–11760.

[5] N. Cennamo, D. Massarotti, R. Galatus, L. Conte, L. Zeni, Performance Comparison of Two Sensors Based on Surface Plasmon Resonance in a Plastic Optical Fiber, Sensors 13, (2013) 721-735.

[6] N. Cennamo, A. Varriale, A. Pennacchio, M. Staiano, D. Massarotti, L. Zeni, S. D'Auria, An innovative plastic optical fiber-based biosensor for new bio/applications. The Case of Celiac Disease, Sens. Actuators B: Chem. 176, (2013) 1008–1014.

[7] N. Cennamo, M. Pesavento, G. D'Agostino, R. Galatus, L. Bibbò, L. Zeni, Detection of trinitrotoluene based on SPR in molecularly imprinted polymer on plastic optical fiber, Proceedings of SPIE 0277-786X, V. 8794, Fifth European Workshop on Optical Fibre Sensors, Krakòw, Poland 19-22 May 2013

[8] Dwivedi, Y.S.; Sharma, A.K.; Gupta, B.D. Influence of design parameters on the performance of a SPR based fiber optic sensor. Plasmonics 2008, 3, 79–86

[9] R. K. Verna, A. K. Sharma, B. D. Gupta, Modeling of Tapered Fiber-Optic Surface Plasmon Resonance Sensor With Enhanced Sensitivity, IEEE Photonics Technology Letters, VOL. 19, NO. 22, (2007) 1786- 1788.

[10] J. Homola, Surface Plasmon Resonance Based Sensors, Springer Series on Chemical Sensors and Biosensors, Springer-Verlag, Berlin-Heidelberg-New York, 2006

[11] A. Abbas, M. J. Linman, Q. Cheng, New trends in instrumental design for surface plasmon resonance-based biosensors, Biosensors and Bioelectronics 26, (2011) 1815–1824.

[12] S. Scarano, M. Mascini, A. P.F. Turner, M. Minunni, Surface plasmon resonance imaging for affinity-based biosensors,Biosensors and Bioelectronics 25, (2010) 957–966.

Chapter 8

Optical Sensors Applied in Agricultural Crops

Fabrício Pinheiro Povh and
Wagner de Paula Gusmão dos Anjos

Additional information is available at the end of the chapter

http://dx.doi.org/10.5772/57145

1. Introduction

There is a very wide range of optical sensors applied in agriculture, which goes from sensors used to analyze soil attributes to sensors installed in combines to measure protein content in wheat grains while they are being harvested. But in this chapter we are going to discuss about optical sensors that, from a short distance are able to measure the agricultural crop's reflectance using specific wavelengths, and how can we use this information.

This kind of optical sensors began to be studied in 1991 with the development of sensors focused in weed detection at the Oklahoma State University. Just based on the simple fact that soil and plants (weeds) have a different interaction with the light emitted from the sensors, allowing identifying what is soil and what is a plant.

In 1992 there was the first discussion between the Departments of Plant and Soil Sciences and Biosystems and Agricultural Engineering concerning the possibility of sensing biomass in wheat and bermudagrass. The objective was to use biomass as an indicator of nutrient need (based on removal). In 1993 Dr. John Solie, Dr. Marvin Stone and Shannon Osbourne collected sensor readings at ongoing bermudagrass, with nitrogen rates versus nitrogen timing experiments with the Noble Foundation in Ardmore, Oklahoma. Initial results were promising enough to continue this work in wheat. And in fall of 1993 the first variable application of nitrogen was done across a 70 meter transect. In 1994 John Ringer and Shannon Osbourne collected sensor readings and later applied variable N fertilizer rates based on the first bermudagrass algorithm developed by TEAM-VRT.

In the subsequence years the research advanced creating algorithms for nitrogen application in many different crops. And nowadays we have commercial sensors being sold to farmers to make real-time application of nitrogen, growth-regulators and desiccants. The objective of this chapter is to show different applications of optical sensors in agricultural crops. At the ABC

Foundation (A private research institution maintained by farmers created in 1984) located in south Brazil, the studies began in 2006 with commercial sensors. Making applied research, the results can be delivered immediately to the agronomists of five cooperatives (Capal, Batavo, Castrolanda, Coopagrícola and Holambra).

The topics to be discussed in this chapter are:

1. The types of sensors used;
2. The range in the spectrum used by sensors and response on crops;
3. Aspects of use in the field and what could affect the measurements;
4. Experiments and the results obtained with sensors;
5. Use of optical sensors combined with a GPS receiver to create maps;
6. Nitrogen application based on sensor's measurements;

2. The types of sensors used

The sensors to measure crops reflectance can be classified according to the platform, like satellites, aerial (airplanes, UAV's - unmanned aerial vehicles, balloon) and ground based. For satellites, airplanes and UAV's it is most common to use cameras to collect images, and ground based optical sensors can collect reflectance data and storage in a text file. The ground sensors can also be classified into active or passive. The basic difference is that the passive sensors need an external source of light, like the sun. The active sensors have its own source of light, which can be a wide range light or a specific wavelength.

There are available a few brands on the market, each one with its own construction characteristics like internal batteries, GPS antenna, data logger and log frequency. The Table 1 has some examples:

Manufacturer	Sensor	Country
AgLeader	OptRx	United States
Falker	ClorofiLOG	Brazil
Force A	Dualex and Multiplex	France
Fritzmeier	ISARIA	Germany
Holland Scientific	CropCircle	United States
Konica Minolta	SPAD	Japan
Topcon	CropSpec	Japan
Trimble	GreenSeeker	United States
Yara	N-Sensor	Germany

Table 1. Optical sensors and manufacturers.

There are also some differences about the distance from the sensor to the crops. For example, the SPAD and the ClorofiLOG sensors need to make static measurements, touching the sensors on crop's leaves. The other sensors make the measurements from centimeters to meters, but do not need to have contact with the leaves.

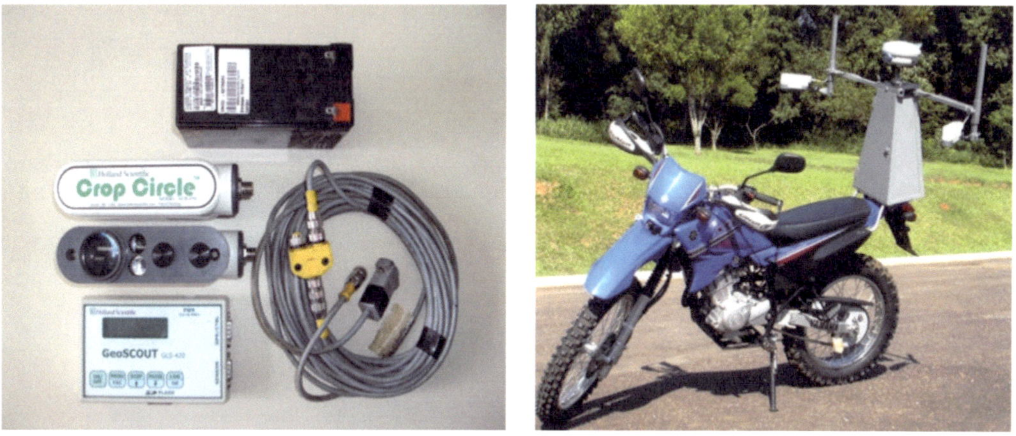

Figure 1. Crop Circle optical sensor.

Let's use one of the available sensors as an example, the Crop Circle® ACS-210 sensor, manufactured by the Holland Scientific Inc., Lincoln, Nebraska. As show in the picture (Figure 1) this sensor has one LED (Light Emitting Diode) that emits active radiation simultaneously in visible and near infrared light with a system called PolySource™ and two silicon photodiodes with a spectral range of 320 to 1,100 nm to detect light. One detector works between 400 and 680 nm and the other between 800 and 1,100 nm. Using a filter on each detector, the wavelengths used are the amber light (590 ± 5 nm) and near infrared (880 ± 10 nm). The sensor must be placed between 0.25 and 2.13 m from the target, and then the light reaches the target and reflects part of the energy, which is received by the detectors. This sensor needs external batteries (12 V) and GPS antenna, but has its own data logger, that saves the data in a SD card. Also in Figure 1 there is an example on how to install the sensors in a motorcycle to collect data on-the-go.

The Figure 2 has two other examples, one is the Hand held GreenSeeker, from Trimble, and the other one is a prototype from Oklahoma State University, which now is also being commercialized by Trimble. The Pocket Sensor was designed for small farmers, made to be a low cost sensor, it does not use GPS signal or a data logger, and you can just position the sensor over the crop, press the trigger and see the value of NDVI (Normalized Difference Vegetation Index).

Figure 2. GreenSeeker sensor collecting NDVI values for weed (a) and (b); detailed pictures from GreenSeeker Hand Held (c) and Pocket Sensor (d); and GreenSeeker mounted on a sprayer (e) and on a motorcycle (f) collecting NDVI values in sugarcane.

3. The range in the spectrum used by sensors and response on crops

The electromagnetic spectrum goes from gamma rays to radio waves. These sensors used for measure crop reflectance usually work in the visible and near infrared region of the spectrum, and combining at least two wavelengths to calculate vegetation indices. From the agronomic point of view, the visible light has a straight relationship with the chlorophyll content, absorbing the blue and red lights, and reflecting the green light. That's what makes us to see healthy plants as green. The near infrared light, not visible by the human eye, is reflected by the mesophyll cells, which is found in more quantity in a plant than chlorophyll, resulting in a much higher reflectance than visible lights. Using both wavelengths it is possible to evaluate the color and biomass production of a crop. In practice, greener and higher biomass plants have a higher chance to have higher yields.

Remote sensing can be defined as the technique of acquisition and application of information about an object without any physical contact. The information is acquired by the detection and measurement of changes that an object imposes to the environment around it, and this signal may include an electromagnetic field emitted and/or reflected, acoustic waves reflected and/or disturbed by the object or the disturbances of gravitational field or magnetic potential with the presence of the object. Usually the acquisition of information is based on capturing the electromagnetic signals that cover the entire spectrum of electromagnetic waves to radio long waves, passing through microwaves, submillimeter, thermal and near infrared, visible, ultraviolet, X ray and gamma ray [1].

The sensors used by remote sensing are devices capable of detect and register electromagnetic radiation in certain range of electromagnetic spectrum and generate information that may be transformed in a product, which can be interpreted, like an image, graphic or tables. The sensor systems are basically formed by an optical part, constructed with lens or mirrors that have the objective of receiving and directing the energy from the targets to the detectors.

The spectral reflectance measurements are the non-destructive approach most promising for determining nitrogen deficiency in crops [2]. For this purpose, remote sensing has been used to evaluate crop conditions related to nitrogen [3]. Many researchers used remote sensing to estimate crop parameters like LAI (Leaf Area Index) [4], leaf chlorophyll content [2], soil cover [5], dry matter [6], water content [7], yield [8], nitrogen content [9] and many others.

In reference [10] studying spectral signature of green leaves, was found that wavelengths from 400 to 700 nm (visible), the reflectance is low, about 10%, with a smooth increase at 550 nm (green). In the near infrared region (700 to 1,300 nm) there is another increase in the reflectance, close to 50%. For the visible light the low reflectance is related to the absorption associated to leaf pigments, mainly chlorophyll. And the reflectance increase at the near infrared region is due the internal structure of the leaves (size and shape of the cells and empty spaces). Combining both visible and near infrared reflectance there are the vegetation indices, which can be resulted from two or more spectral bands.

First, [11] it was proposed the ratio between the measurements from 800 and 675 nm to determine the leaf area index in forests. The relation between these two wavelengths is known as RVI (Ratio Vegetation Index). The NDVI (Normalized Difference Vegetation Index) appeared just after [12], which was found a relation between two wavelengths that better solved the issues about soil interference on the vegetation measurements, and also reduced the atmosphere influence and sun angle variations.

The normalization proposed guarantees that the values obtained from the NDVI are contained in the same scale of values, between -1 and 1, as shown in the equation (1).

$$NDVI = \frac{(\rho_{IR} - \rho_V)}{(\rho_{IR} + \rho_V)} \tag{1}$$

Which:

ϱ_{IR} = infrared reflectance;

ϱ_V = visible reflectance.

The normalization is produced by the combination of the strong absorption by the chlorophyll in the red region and the strong reflectance in the near infrared, due to the dispersion in the leaf mesophyll and the absence of absorption by the pigments [13]. The peculiarity assigned to the NDVI is the early saturation, what makes it insensible to biomass increase after certain development stage. That means the NDVI stabilizes, showing constant values, even with the biomass increase [10].

4. Aspects of use in the field and what can affect the measurements

Measurements in the field can be affected by the sensor positioning, like the distance from the crop, dependence of a light source, the presence of dew over the leaves and also because of factors that can stress the plants. The light source classifies the sensors in active and passive sensors. The passive sensors are dependent of sunlight, not working at night or might show different readings when there are clouds or shadows. There is a limitation of the distance from the target to the sensor, because if it is too close the sensor may not capture the reflectance, and if it is too far the data may have noise signals. And the presence of dew is just because the presence of water over the leaves can change the reflectance in both visible and near infrared. With the presence of dew the reflectance increases, but as visible light is more affected, consequently reduces the NDVI values.

Other methods can be used instead of on-the-go sensing to indirectly measure nitrogen stress, like chlorophyll content in the leaves using chlorophyll meters [14] like SPAD and ClorofiLOG. In reference [15], analyzing aerial and satellite images it was found lower correlation with wheat crops variables then ground sensors, besides depending on the weather it is not possible to have satellite images due the presence of clouds.

The spectral data that can be obtained from satellite images have low temporal resolution to use in agriculture. The ground acquisition is independent from the weather, the data collection can be done together with other machinery operation and the data are available just after you finish, with no need of complicated processing [16]. Higher correlation was found between the NDVI obtained from the GreenSeeker then the NDVI obtained from the satellite Quickbird II images for the amount of nitrogen applied, nitrogen content on the flag leaf, yield and protein content in wheat grains [15].

In reference [17] the author used a sensor installed in the front of a tractor to acquire spectral data in wheat. The tractor was at the speed of 0.8 m/s, and the sensor was set to register data with the frequency of 10 measurements per second. The authors found good relationship between the reflectance and nitrogen absorption by the wheat crop. The same configuration was used to collect data in pasture and also found good relationship between the NDVI and the nitrogen removal by the crop [18]. But depending on the sensor used, the speed can also affect the NDVI, increasing the coefficient of variation of the measurements.

An experiment conducted along the day, collecting NDVI values at the same spot but in different times of the day showed that the presence of dew on the leaves reduced the NDVI values of 12% for GreenSeeker and 27% for Crop Circle, from the first (7:30 am) to the last reading (11:30 am). Near infrared showed small differences in the readings, not significant, unlike the results of the visible data, but they both increased. These results lead to conclude that the visible wavelength is more affected by the presence of dew than the near infrared. Then, the presence of dew should be considered for nitrogen recommendation based on active optical sensors.

Stressed plants can show a decrease in the absorption by the chlorophyll, also decreasing the reflectance in the near infrared due to changes in the cells structure [19]. And that is what makes the sensors to be promising while evaluating nitrogen stress. Many studies were realized to estimate nitrogen deficiency in corn [3], wheat [20], edible beans [21], cotton [22], citrus [23], barley [24] and sugarcane [25]. These authors showed great potential to use sensors to estimate nitrogen content in crops.

5. Experiments and results obtained with sensors

As discussed previously, the sensors are able to measure differences in color and biomass production. In the field when nitrogen is applied and the plant absorbs it, the result is a plant greener and with more biomass. The Figure 3 shows the first experiment realized by ABC Foundation in Brazil in the winter of 2006 with the result of nitrogen application in wheat and the effect on the NDVI measured by the GreenSeeker sensor in two different soils.

The graphic in Figure 3 is a typical result of nutrient application in agricultural crops, with a quadratic model. There is always a limit, technical or economic, which beyond that there is no gain in yield. If linear models explained fertilizers application, where more fertilizer applied should result in more yield, it would be much easier, but there is also a physiological response, when higher nutrients rates become toxic to the plant. This result shows that applying more nitrogen we could find a response from the NDVI measured with a GreenSeeker sensor. In

Table 2 it is possible to see other variables that had significant correlation with the NDVI in the same experiment.

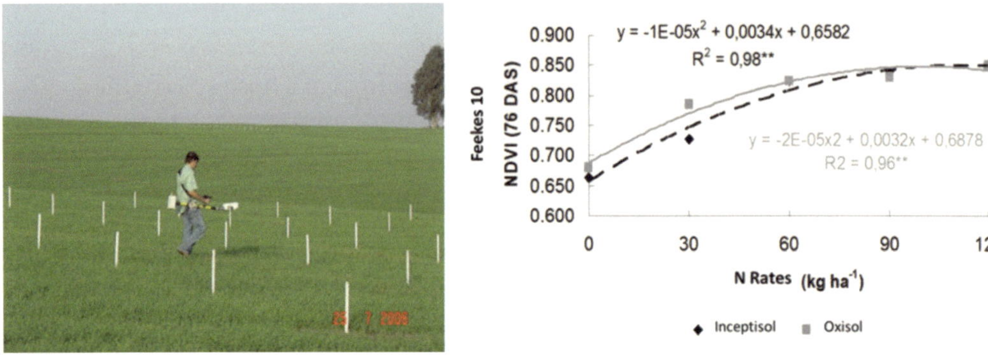

Figure 3. First experiment and NDVI response with nitrogen rates in wheat.

Variables	Inceptsol	Oxisol
	(r)	
Leaf N x Yield	0,99**	0,95**
Dry Matter x Yield	0,98**	0,97**
NDVI x Leaf N	0,95**	0,88*
NDVI x Dry Matter	0,97**	0,99**
NDVI x Yield	0,95**	0,98**

Table 2. Coefficients of correlation (r) between the evaluated variables in a wheat experiment in two different soil types. Leaf nitrogen content, dry matter and NDVI were evaluated 76 days after sowing, at the Feekes 10 stage of development.

6. Use of optical sensors combined with a GPS receiver to create maps

Maybe one of the most interesting applications for optical sensors in agriculture is to be able to use geographical coordinates to create maps from reflectance measurements. But why do that? Precision Agriculture is based on the fact that all fields have variability. Understanding this variability is first step to make decisions about investments in Precision Agriculture. Many procedures may be used to characterize and treat spatial variability on yield aiming profit for the farmers, but a wide and safe vision about the impact of the variability in a production system requires an accurate measurement of this variability.

Soil variability is caused by climate, topography, vegetation, soil forming processes and also management. These factors can influence the variability in different scales and cause great variability on nutrient availability in the soil. Then, when using uniform rates of fertilizers it is almost certain that excessive rates will be applied to some parts of the field and inadequate rates in others.

Precision Agriculture can be defined as a management system that considers spatial variability that is present in a production field, independently of the size and treats locally this variability. It is well proved that quality and yield are spatially variable and systems are being developed to explore these variations and increase profit [26]. The variable rate application of fertilizers is one of the options to manage variability, and creating maps from optical sensors can help to realize variable applications with nitrogen.

An optical sensor connected to a GNSS (Global Navigation Satellite System) receiver is able to register reflectance values with a pair of coordinates (latitude and longitude), and when the file is imported to a GIS software (Geographic Information System) it is possible to represent the measurements and its distribution in a field with maps (Figure 4).

Depending on sensor system used, that information can be stored in different ways and types of files. Some systems installed in agricultural machinery allow using a laptop with software that will store the sensor readings and coordinates. Other systems use their own data logger or a pocket pc. The Figure 4 is an example of a raw data collected in a field. Each strip is around 24 meters from each other, but what looks like a line, actually are many dots very close to each other due the frequency used to collect the data in the field. If we look closer, we see as in Figure 5.

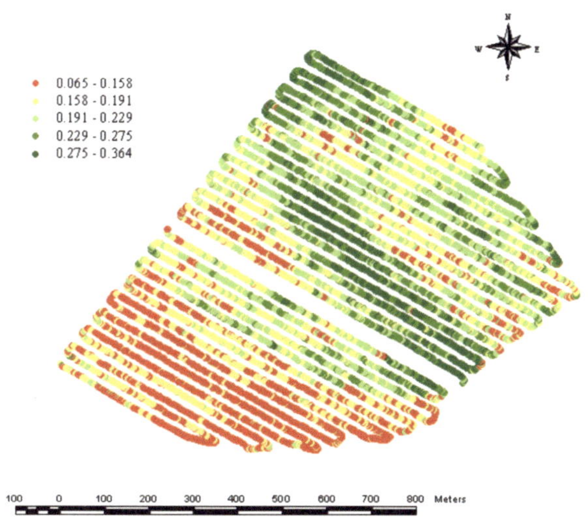

Figure 4. NDVI raw data.

As showed before in the picture of the motorcycle with one sensor on each side, when looking closer at the data we can see that in every coordinate we have a NDVI value from each sensor.

Figure 5. NDVI raw data (zoom in).

As the frequency is very high, lets say 5 Hz or 5 measurements per second, if the motorcycle moves at 5 m/s (18 km/h) with 24 m distant from the next pass, the sensor registers two NDVI values every meter and more than 800 measurements per hectare. But the raw data may have noise, so after applying some statistical filters to remove outliers and using some kind of interpolation method, for example the inverse distance, it is possible to create a surface map with a regular grid that will look like a raster image. The Figure 6 is a NDVI map after interpolating the raw data from a corn field.

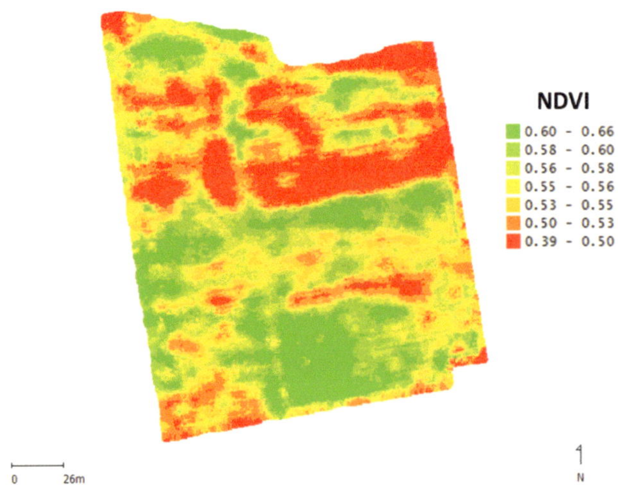

Figure 6. Interpolated NDVI map of a 2.86 ha field with a corn crop.

http://dx.doi.org/10.5772/57145

Figure 7. Picture of corn plants with low NDVI values.

This maps shows that in the red areas the corn plants had a low development and because of that will have lower yields (less than 10 tons/ha), while the green areas had yields over 12 tons/ha. This information is important for precision agriculture users that are interested in finding the specific problem for that particular location. Looking at plants in the field, right in middle of the map in the red area, we can see the plants as showed in Figure 7.

The NDVI map created while the corn crop is still at the V6 stage of development, allow the farmers and agronomists to go back in the field and see what is causing this slower growth of the plants. And as the information is stored, it is possible to go back every year at the same spot for comparison, collect soil and plant samples, send to the lab and try to understand what the problem is. That means the farmer can treat each part of the field accordingly with its own characteristics.

7. Nitrogen application based on sensor's measurements

Why manage nitrogen in agricultural crops? Because some crops do not have the ability to fix atmospheric nitrogen like soybeans do, so nitrogen is applied to crops from mineral fertilizers.

Among the nutrients, nitrogen (N) is the most important and essential for crop development and is also important from an environmental perspective. Nitrogen is a constituent of chlorophyll, the first pigment to absorb the light energy necessary for photosynthesis. Plants usually have 1 to 5% nitrogen in their tissues, and if used appropriately with the other nutrients, N addition can accelerate the development of corn and other grains [27]. And unlike other important nutrients as phosphorus or potassium that are measured in the soil with chemical analyses of soil samples, nitrogen is recommended based on historical response curves.

Plants with N deficiency have yellow leaves and reduced growth. Not only does the absence of N limit yield, but so can excess N. In most cases, N application recommendations are based on average conditions, and farmers do not use the most advantageous fertilizer combinations, even when there is no financial limitation [28].

The absorption of N by crops is variable among and between seasons, as well as between locations in the same field, even when the N supplies are high [29]. The N supply from soil to crop varies spatially. Consequently, the demand for N by the crop also varies. As a result, the crop's nutritional status is a good indicator of the necessary N rate application [30].

Traditional N management around the world is generally inefficient, creates environmental contamination and is controversial. To provide appropriate recommendations of spatial N applications, it is necessary to use several tools simultaneously, such as crop and soil sensors, to achieve reliable measurements of N availability from soil and crops needs [31].

The evaluation of nitrogen use efficiency (NUE) in agriculture is an important way to evaluate the density of N applied and its role in yield [32]. Because crop responses to N application depend on the organic matter in the soil, strategies of N management in cereal crops that include reliable predictions of the response index in each season could increase NUE [33].

In this scenario, sensors are becoming more prevalent in agricultural lands. Using variable rate equipment, it is possible to detect variability in crops and make rapid decisions in the field. Some sensors allow real time changes in agricultural practices by detecting variability and responding to that variability [34].

In reference [20] they developed a methodology to apply nitrogen based on crops reflectance, using a yield parameter sensible to local conditions, intrinsic and that could reflect yield potential, possible to be used in season. Unlike other models that need several parameters to predict plant growth, the optical sensors use the plant as an indicator. Later [35] they created an index called INSEY (in-season estimated yield), which is calculated dividing the NDVI value by the number of days from sowing to sensing. The relation between the INSEY and Yield generate an exponential model using plot experiments, which can be used to estimate yield just based on the sensor measurements. Estimating yield where there is and where there is not nitrogen, by the difference in yield it is possible to calculate the nitrogen rate to make them to have the same yield.

This method considers the spatial variability of yield potential, the absorption of nitrogen applied at sowing and the crop's response to an additional N rate. NUE were increase in 15%

and showed that the measurement of crop's reflectance can be used to calculate more efficient and profitable N rates [35]. The ability to identify areas where the crop will respond to the fertilizer applied is important, if the response to N is expected, then management strategies can be modified [36].

Many field experiments were realized in order to create the algorithms that convert the sensor readings in nitrogen rates and to validate the results. Depending on the year, savings of none to 75% of nitrogen were found. For example the Figure 8 shows the NDVI and yield response to nitrogen rates. Using the 120 kg of nitrogen per hectare as the standard rate, the NDVI measured 50 days before the harvest indicates that the wheat crop won't respond to rates over than 60 kg of nitrogen per hectare. And the wheat yield obtained after the harvest shows that rates over than 60 kg/ha will affect yield negatively. Therefore, in this example, the nitrogen rate was 50% lower than the commonly recommended rate.

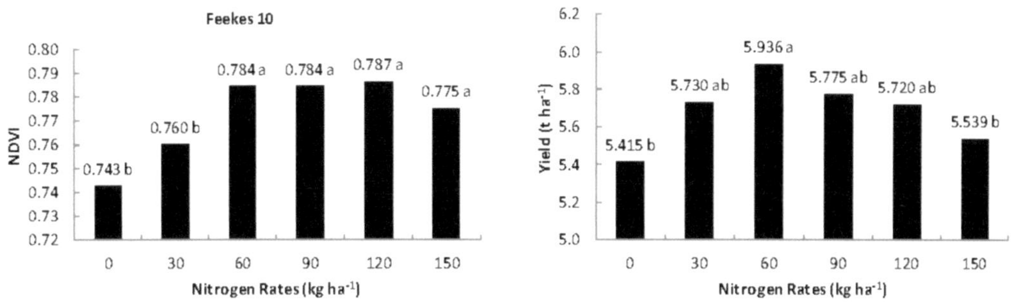

Figure 8. NDVI and wheat yield response with nitrogen rates applied.

The yield goal, often used for nitrogen recommendations, is the yield that the farmer expects to produce. But what you expect to produce and what really is going to be produced are usually different. In [20] they exemplify that the North Dakota State University recommend using the maximum yield obtained in the last 5 years as yield goal, because it is usually 30% higher than the average. The yield goal can also be calculated adding 5 to 10% to the average yield from the last 5 to 7 years [37]. One example is to use 25 kg of nitrogen per ton of corn expected. But management practices and the weather have a huge influence on yield. Climate can vary significantly from one year to another, what may cause great differences in yield potential [31].

Now let's use an experiment as an example of how to realize a nitrogen recommendation using variable rate across a field. The experiment was realized in South Brazil, Paraná State, farm Manzanilha (25° 28′ 55″ S, 50° 21′ 07″ W), during 2007 winter. The sensor used was the GreenSeeker Hand Held^{TM} measuring the NDVI. The treatments (Table 3) were disposed in strips with fixed rate of nitrogen (6 x 600 m) and variable rate of nitrogen (11 x 600 m), with four treatments and four replications. The wheat was sowed in May 28th. The Sketch is presented in Figure 9.

Treatments	N Sowing	N Feekes 1	N Feekes 5	N Feekes 10	Total N
1	18,4	0	0	Sensor	18,4
2	18,4	34	0	Sensor	52,4
3	18,4	0	102	0	120
4	18,4	34	68	0	120

Table 3. Treatments with nitrogen rates (kg/ha)

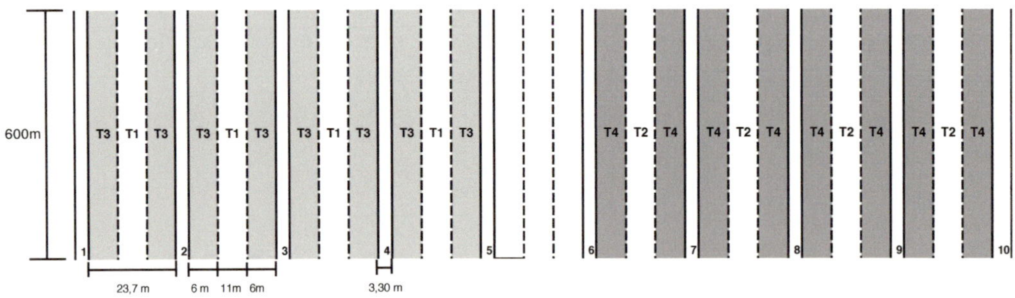

Figure 9. Experiment sketch.

All treatments received 18.4 kg/ha of nitrogen at the sowing; treatments 2 and 4 had an extra 34 kg/ha just after sowing and treatments 3 and 4 received more 102 and 68 kg/ha of N, respectively, on Feekes 5 to complete the 120 kg/ha of N, recommended based on historical response curves. Treatments 3 and 4 were used as control, and treatments 1 and 2 received variable rates of nitrogen at Feekes 10 based on the sensor readings.

The measurements were realized with the GreenSeeker sensor mounted on a motorcycle 79 days after sowing (Feekes 10). The pictures of the field and data collection can be seen in Figure 10.

Figure 10. Details of the sensor mounted on the motorcycle (a) and data collection (b).

After the measurements, the data were imported to a GIS software (SSToolBox) to create a map with the NDVI raw data (Figure 11).

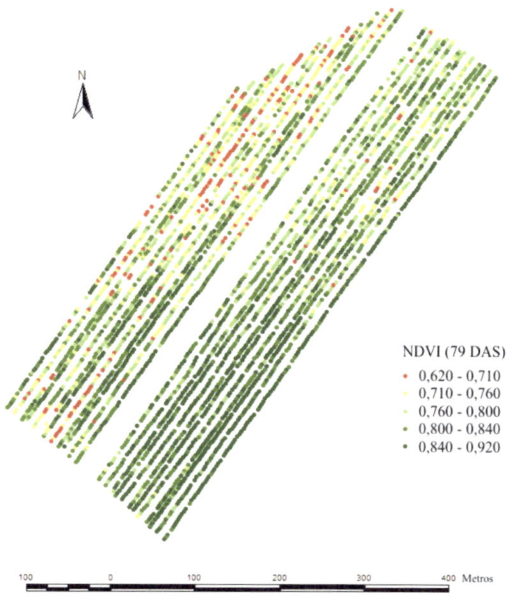

Figure 11. NDVI raw data in wheat.

Looking at the NDVI map it is possible to notice that even at the 120 kg/ha of N strips there was spatial variability, so it was decided to adapt the methodology proposed in [35], and do not use the average of the strip as a control, but divide the strip in three levels according with the topography. The elevation map (Figure 12) was created from the data collected by the GPS receiver. It was used three NDVI averages for treatment 3 and more three averages for treatment 4, with a total of six NDVI values to be used as a reference and estimate yield.

After dividing the elevation in three ranges it was calculated the average of the NDVI values for the 6 regions. Then the response index (RI) was calculated (eq. 2) with the ratio between the average NDVI from strips with 120 kg/ha of N and the NDVI from every point in the map. That means that every point has an NDVI value and a RI value.

$$RI = NDVI_{rich\ strip} / NDVI_{field} \quad (2)$$

Which:

$NDVI_{rich\ strip}$ = NDVI from 120 kg/ha of N strips;

$NDVI_{field}$ = NDVI from all points in the field that will be applied N;

RI = response index.

The response index means that if the number is higher than 1, the NDVI from the strip with more N is higher than the NDVI from the parts of the field that will be applied, so the plants are using the N. And if RI = 1 means that where there is N and where there is no N the NDVI is the same, so there is no need to apply more N. Using RI it is possible to identify the parts of the field that will need more nitrogen, and save where there is no response, because it will not have increase in yield.

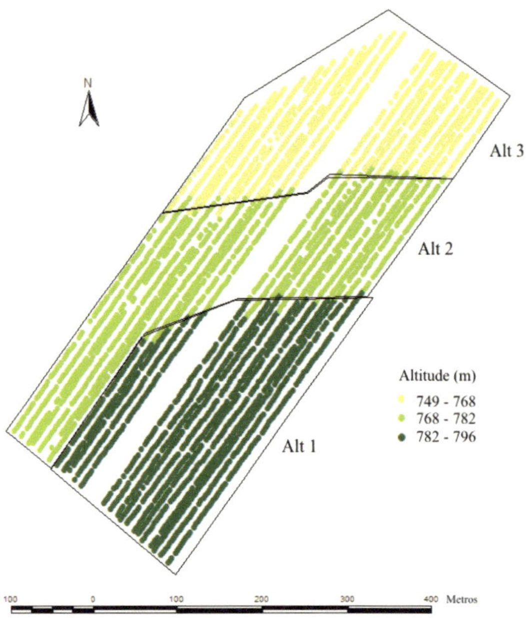

Figure 12. Elevation map divided in three levels.

To calculate the N rate was used the methodology from [35], identifying the difference of yield between the N rich strip and the rest of the field, and then applying the correct N rate to reach the same yield. That means the objective is to save nitrogen and not increase yield. The N rates calculated were used to create a nitrogen application map (Figure 13).

The calculated rates varied from 0 to 60 kg/ha of N, but they were simplified into three ranges and applied the maximum of each range. For example, the rate 20 kg/ha represents the rates from 0 to 20 kg/ha and so on. This simplification was done to make the application easier. To apply the nitrogen it was used a sprayer mounted on a tractor for liquid nitrogen application (Figure 14). To vary the rates along the field the tractor changed the speed, for each one of the rates was determined a specific gear of the tractor and a specific engine rotation. The sprayer was calibrated to apply 20 kg/ha (2.5 m/s), 40 kg/ha (1.25 m/s) and 60 kg/ha (0.8 m/s). The gears and the engine rotation for tractor were determined with a 50 m space measuring the time that the tractor took to move the 50 m. The sketch of the application is in Figure 15.

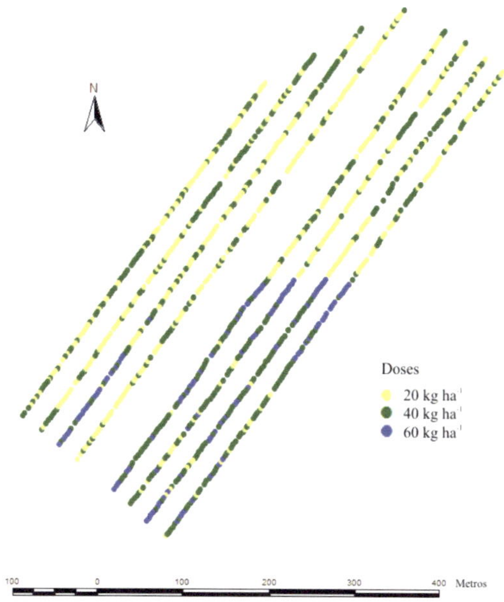

Figure 13. Nitrogen map from the calculated rates.

Figure 14. Pictures of the liquid nitrogen fertilizer application.

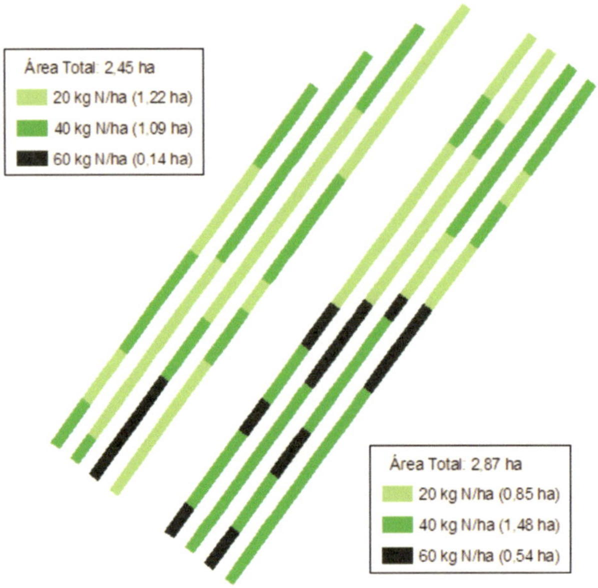

Figure 15. Sketch of the nitrogen application.

The harvest of the strips was realized using a combine John Deere 9650 STS with a 30 feet header and equipped with a yield monitor AgLeader PF3000 Advantage, what made possible to generate the yield map. The Figure 16 shows the yield map that was similar to the NDVI map. We can see in the map that each strip has a pair, because one strip is the 120 kg/ha of N and the other is the variable rate application. The yield varied from 1,500 to 5,500 kg/ha of wheat. But even the strips with 120 kg/ha of N had some parts of the field with lower yields, which means that these regions of low yield have some other problem and not nitrogen deficiency. Comparing with the application sketch the region of higher rates are same regions of high yields.

The left side of the experiment (first 4 pair of strips) had an average yield of 3,053 kg/ha using variable rate application, with 69.3% of saving in nitrogen, and 3,026 kg/ha using 120 kg/ha of N. The right side of the experiment had an average yield of 3,568 kg/ha using variable rate, with 42.5% of saving in nitrogen, and 3,546 kg/ha using 120 kg/ha of N. There was no statistical difference between the yield of fixed rate and variable rate, but the savings were high. The saving of nitrogen was higher in the region with lower yield potential.

As high is the yield goal just with the nitrogen supplied by the soil, with no additional N (low RI), in general, lower will be the N rates to reach maximum yields. If the RI for a field is low (RI < 1.1), that means the places with no N applied are similar with the places with N, and probably the response to an additional application will also be low. But if the RI is high (RI > 1.1), probably the crop will respond to the fertilization, so complementary rates should be applied [36].

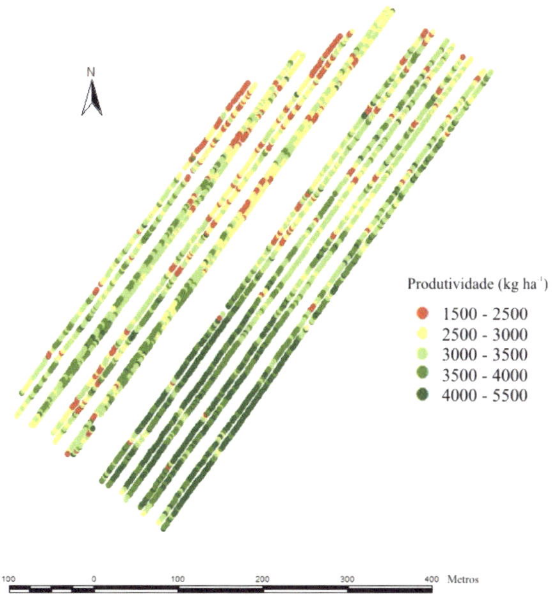

Figure 16. Yield map collected by the John Deere combine.

Ps: In each pair of strips, the left one is the 120 kg/ha of N and the right one is the variable rate application.

There is NDVI spatial variability even in the regions where fixed rates of nitrogen were applied, showing that crops respond in different ways inside the same field. The methodology used to apply nitrogen in variable rate, using the crop as an indicator and optical sensors to measure is really promising, being able to save nitrogen in places where there is less use by the plants.

However, optical sensors do not consider yield potential. Reflected light by the crop in a high yield zone can show an appropriate content of N at the moment of measurement, but N can still be deficient before maturation. In a low yielding zone, reflected light may suggest a higher need for N, inducing an excess rate application for plants that are limited by other factors [38].

Large differences were found between two corn fields, one with yields <12,000 kg/ha and the other with yields >17,000 kg/ha [39]. These authors suggested that considering climate conditions between planting date and the measurements could lead to a better relationship between yield estimation and actual yield because climate varies among areas and among years. Even when data from different areas and years are combined, the resulting model may not be reliable due spatial and temporal differences. These differences are some of the primary problems in developing a model that can be used widely to estimate yield by optical sensors. They also highlight that there is much room for improvement.

Variables that affect N availability were included, such as organic matter and soil exchange capacity, and variables that are related to N demand by the crop, such as solar radiation [40].

Additionally, the variability of rain distribution and water stored in soil at the time of planting can affect yield and management decisions [41].

Based on the observations of long-term experiments, it was found that the treatments with higher yield did not always correspond to higher N application rates. In some years, maximum yields were observed with low N application rates. In 3 long-terms experiment, there was no relationship between RI and yield. These results indicated that RI and yield are independent, and so they need to be estimated separately. Any practice that does not consider the independence of RI and yield can lead to misinformed recommendations. If yield potential varies among years along with N demand, the obvious solution to the increased efficiency of N recommendations is to be able to estimate crop response to N and yield potential [42].

8. Conclusions

The NDVI obtained from optical sensors are data that look very promising to map and analyze spatial variability of crop's development and for nitrogen management. It is possible to find significant regressions between NDVI and many crops' parameters. The technology of using optical sensors to recommend nitrogen showed high savings of the nutrient, making agriculture more profitable and with lower risk of water contamination by nitrate. Research with the application of optical sensors in agriculture is still scarce and need to expand to a wide range of conditions, for many other crops and applications.

Author details

Fabrício Pinheiro Povh and Wagner de Paula Gusmão dos Anjos

*Address all correspondence to: fppovh@gmail.com; fabricio@fundacaoabc.org.br

Fundação ABC, Agricultural Machinery and Precision Agriculture Department, Castro, Brazil

References

[1] Liu WTH. Remote Sensing Application (in Portuguese). Campo Grande: Uniderp; 2007.

[2] Tumbo SD, Wagner DG, Heinemann PH. Hyperspectral characteristics of corn plants under different chlorophyll levels. Transactions of the ASAE 2002;45(3) 815-823.

[3] Blackmer TM, Schepers JS, Varvel GE, Walter-Shea EA. Nitrogen deficiency detection using reflected shortwave radiation from irrigated corn canopies. Agronomy Journal 1996;88(1) 1-5.

[4] Baret F, Guyot G. Potentials and limits of vegetation indices for LAI and APAR assessment. Remote Sensing of Environment 1991;35(1) 161-173.

[5] Boissard P, Pointel JG, Tranchefort J. Estimation of the ground cover ratio of a wheat canopy using radiometry. International Journal of Remote Sensing 1992;13(9) 1681-1692.

[6] Tucker CJ, Holben BN, Elgin JH Jr., McMurtrey, J.E. Remote sensing of total dry-matter accumulation in winter wheat. Remote Sensing of Environment 1981;11(1) 171-189.

[7] Waheed T, Bonnell RB, Prascher SO, Paulet E. Measuring performance in precision agriculture: CART – A decision tree approach. Agricultural Water Management 2006;84(1) 173-185.

[8] Fischer RA, Howe GN, Ibrahim Z. Irrigated spring wheat and timing and amount of nitrogen fertilizer: I - Grain yield and protein content. Field Crops Research 1993;33(1) 37-56.

[9] Solie JB, Stone ML, Raun WR, Johnson GV, Freeman K, Mullen R, Needham DE, Reed S, Washmon CN. Real-time sensing and N fertilization with a field scale GreenSeeker applicator: Proceedings of the International Conference on Precision Agriculture, Minneapolis, 2002.

[10] Moreira RC. Influência do posicionamento e da largura de bandas de sensores remotos e dos efeitos atmosféricos na determinação de índices de vegetação. Master's dissertation. Instituto Nacional de Pesquisas Espaciais; 2000.

[11] Jordan CF. Derivation of leaf area index from quality of light on the forest floor. Ecology 1969;50(4) 663-666.

[12] Rouse JW, Haas RH, Schell JA, Deering DW. Monitoring vegetation systems in the great plains with ERTS: Proceedings of ERTS Symposium, Washington, DC, 1973.

[13] Woolley JT. Reflectance and transmittance of light by leaves. Plant Physiology 1971;47(5) 656–662.

[14] Piekielek WP, Fox RH, Toth JD, Macneal KE. Use of a chlorophyll meter at the early dent stage of corn to evaluate nitrogen sufficiency. Agronomy Journal 1995;87(3) 403-408.

[15] Wright DL, Rasmussen VP, Ramsey RD, Baker DJ., Ellsworth JW. Canopy reflectance estimation of wheat nitrogen content for grain protein management. GIScience and Remote Sensing 2004;41(4) 287-300.

[16] Steven MD. Correcting the effects of field of view and varying illumination in spectral measurements of crops. Precision Agriculture 2004;5(1) 55-72.

[17] Stone ML, Solie JB, Raun WR, Whitney RW, Taylor SL, Ringer JD. Use of spectral radiance for correcting in–season fertilizer nitrogen deficiencies in winter wheat. Transactions of the ASAE 1996;39(5) 1623–1631.

[18] Taylor SL, Raun WR, Solie JB, Johnson GV, Stone ML, Whitney RW. Use of spectral radiance for correcting nitrogen deficiencies and estimating soil test variability in an established Bermuda grass pasture. Journal of Plant Nutrition 1998;21(11) 2287–2302.

[19] Ayala-Silva T, Beyl CA. Changes in spectral reflectance of wheat leaves in response to specific macronutrient deficiency. Advances in Space Research 2005;35(2) 305-317.

[20] Raun WR, Solie JB, Johnson GV, Stone ML, Lukina EV, Thomason WE, Schepers JS. In-season prediction of potential grain yield in winter wheat using canopy reflectance. Agronomy Journal 2001;93(1) 131-138.

[21] Thai CN, Evans MD, Deng X, Theisen AF. Visible & NIR imaging of bush beans grown under different nitrogen treatments. ASAE 1998;Paper 98-3074.

[22] Sui R, Wilkerson JB, Hart WE, Wilhelm LR, Howard DD. Multi-spectral sensor for detection of nitrogen status in cotton. Applied Engineering in Agriculture 2005;21(2) 167-172.

[23] Min M, LEE WS. Determination of significant wavelengths and prediction on nitrogen content for citrus. Transactions of the ASAE 2005;48(2) 455-461.

[24] Kim Y, Evans RG, Waddell J. Evaluation of in-field optical sensor for nitrogen assessment of barley in two irrigation systems. ASAE 2005;Paper, PNW05-1004.

[25] Frasson FR, Molin JP, Povh FP, Salvi JV. Temporal behavior of NDVI measured with an active optical sensor for different varieties of sugarcane. Revista Brasileira de Engenharia de Biossistemas 2007;1(1) 237-244.

[26] Stafford JV. An investigation into the within-field spatial variability of grain quality: Proceedings of the European Conference on Precision Agriculture, Odense, 1999.

[27] Tisdale SL, Nelson WL, Beaton JD, Halvin JL. Soil fertility and fertilizers. New York: Macmillan; 1993.

[28] Raij B van. Soil fertility and fertilization (in Portuguese). São Paulo: Agronômica Ceres; 1991.

[29] Gastal F, Lemaire G. N uptake and distribution in crops: an agronomical and ecophysiological perspective. Journal of Experimental Botany 2002;53(370) 789-799.

[30] Schächtl J, Huber G, Maidl FX, Sticksel E, Schulz E, Haschberger P. Laser-induced chlorophyll fluorescence measurements for detecting the nitrogen status of wheat (Triticum aestivum L.) canopies. Precision Agriculture 2005;6(1) 143-156.

[31] Solari F. Developing a crop based strategy for on-the-go nitrogen management in irrigated cornfields. PhD thesis. University of Nebraska; 2006.

[32] Fageria NK, Baligar VC. Enhancing nitrogen use efficiency in crop plants. Advances in Agronomy 2005;88(1) 97-185.

[33] Johnson GV, Raun WR. Nitrogen response index as a guide to fertilizer management. Journal of Plant Nutrition 2003;26(2) 249-262.

[34] Inman D, Khosla R, Westfall DG, Reich R. Nitrogen uptake across site specific management zones in irrigated corn production systems. Agronomy Journal 2005;97(1) 169-176.

[35] Raun WR, Solie JB, Johnson GV, Stone ML, Mullen RW, Freeman KW, Thomason WE, Lukina EV. Improving nitrogen use efficiency in cereal grain production with optical sensing and variable rate application. Agronomy Journal 2002;94(4) 815-820.

[36] Mullen RW, Freeman KW, Raun WR, Johnson JV, Stone ML, Solie JB. Identifying an In-Season Response Index and the Potential to Increase Wheat Yield with Nitrogen. Agronomy Journal 2003;95(2) 347-351.

[37] Rice CW, Havlin JL. Integrating mineralizable nitrogen indices into fertilizer nitrogen recommendations. In: Havlin JL, Jacobsen JL. Soil testing: Prospects for improving nutrient recommendations. Madison: Soil Science Society of America; 1994. p.1-14.

[38] Lowenberg-DeBouer J. The Management Time Economics of On-the-go Sensing for Nitrogen Application, SSMC Newsletter, May 2004, Purdue University: http://www.agriculture.purdue.edu/ssmc/Frames/SSMC_May_2004_newsletter.pdf (accessed 5 July 2013).

[39] Inman D, Khosla R, Reich RM, Westfall DG. Active remote sensing and grain yield in irrigated maize. Precision Agriculture 2007;8(4-5) 241-252.

[40] Liu Y, Swinton SM, Miller NR. In site-specific yield response consistent over time? Does it pay? American Journal of Agricultural Economics 2006;88(2) 471-483.

[41] Moeller C, Asseng S, Berger J, Milroy SP. Plant available soil water at sowing in Mediterranean environments—Is it a useful criterion to aid nitrogen fertilizer and sowing decisions? Field Crop Research 2009;114(1) 127-136.

[42] Raun WR, Solie JB, Stone ML. Independence of yield potential and crop nitrogen response. Precision Agriculture 2011;12(4) 508-518.

Chapter 9

Nanostructured Detector Technology for Optical Sensing Applications

Ashok K. Sood, Nibir K. Dhar, Dennis L. Polla,
Madan Dubey and Priyalal Wijewarnasuriya

Additional information is available at the end of the chapter

http://dx.doi.org/10.5772/58349

1. Introduction

This Chapter covers recent advances in nanostructured based detector technology, materials and devices for optical sensing applications. The authors have many years of experience working nanotechnologies that include a variety of semiconductors and other advanced materials such as GaN, ZnO, Si/SiGe, CNT and Graphene for optical sensing applications.

Optical sensing technology is critical for defense and commercial applications including optical communication. Advances in optoelectronics materials in the UV, Visible and Infrared, using nanostructures, and use of novel materials such as CNT and Graphene have opened doors for new approaches to apply device design methodology that are expected to offer enhanced performance and low cost optical sensors in a wide range of applications.

We will cover the UV band (200-400 nm) and address some of the recent advances in nano-structures growth and characterization using GaN/AlGaN, ZnO/MgZnO based technologies and their applications. We will also discuss nanostructure based Si/SiGe technologies (400-1700 nm) that will cover various bands of interest in visible-near infrared for detection and optical communication applications. The chapter will also discuss some of the theoretical and experimental results in these detector technologies.

Recent advancements in design and development of CNT based detection technologies have shown promise for optical sensor applications. We will present theoretical and experimental results on these device and their potential applications in various bands of interest.

2. UV, Visible and infrared spectrum and bands of interest

The Ultraviolet spectrum has been of interest for a variety of sensors for defense and commercial applications. The UV band is from 250-400 nanometers as shown in the figure 1. This band can be further divided into UVA and UVB bands. Each of these bands has applications for sensors, detectors and LED applications.

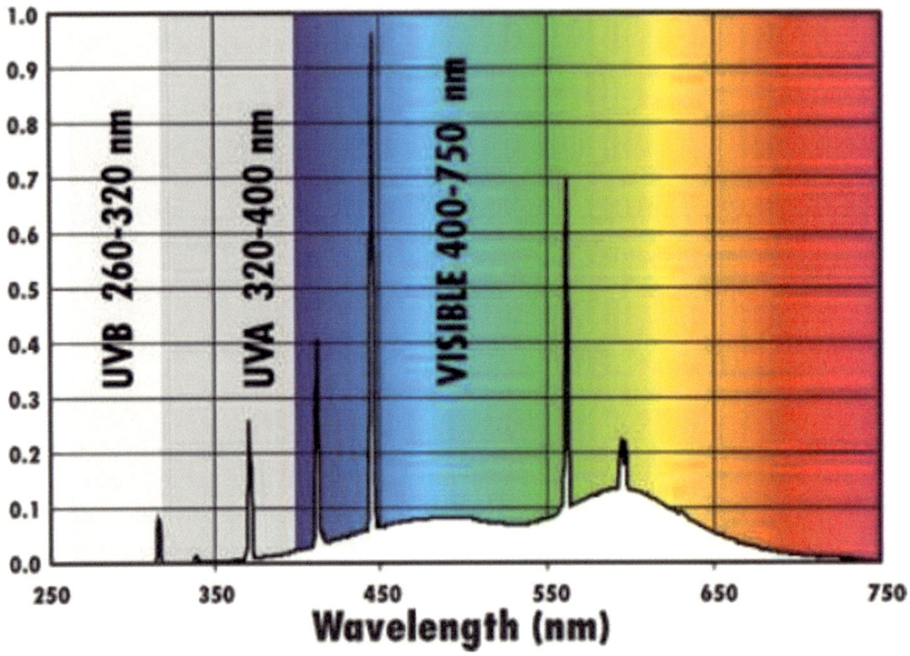

Figure 1. Overview of UV and Visible Spectral Band [1]

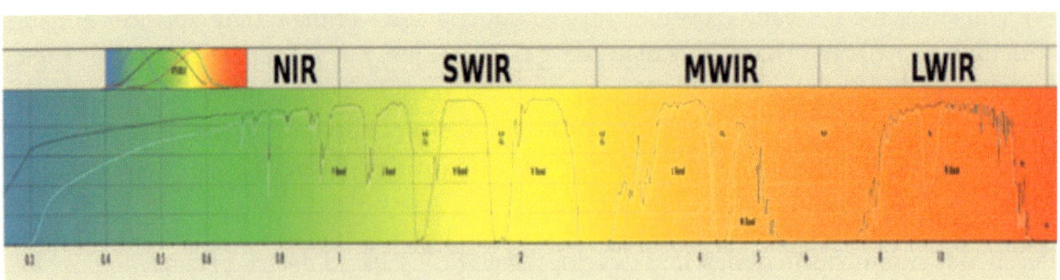

Figure 2. Definition of IR Spectral Band [1].

http://dx.doi.org/10.5772/58349

The word "infrared" refers to a broad portion of the electromagnetic spectrum that spans a wavelength range from 1.0 um to beyond 30 um everything between visible light and microwave radiation. Much of the infrared spectrum is not useful for ground- or sea-based imaging because the radiation is blocked by the atmosphere. The remaining portions of the spectrum are often called "atmospheric transmission windows," and define the infrared bands that are usable on Earth. The infrared spectrum is loosely segmented into near infrared (NIR, 0.8-1.1um), short wave infrared (SWIR, 0.9-2.5um), mid wave infrared (MWIR, 3-5um), long wave infrared (LWIR, 8-14um), very long wave infrared (VLWIR, 12- 25um) and far infrared (FIR, > 25um), as shown in Figure 2. The MWIR- LWIR wavebands are important for the imaging of objects that emit thermal radiation, while the NIR-SWIR bands are good for imaging scenes that reflect light, similar to visible light.

Since NIR and SWIR are so near to the visible bands, their behavior is similar to the more familiar visible light. Energy in these bands must be reflected from the scene in order to produce good imagery, which means that there must be some external illumination source. Both NIR and SWIR imaging systems can take advantage of sunlight, moonlight, starlight, and an atmospheric phenomenon called "nightglow," but typically require some type of artificial illumination at night. In lieu of photon starved scenes, arrays of infrared Light Emitting Diodes (LEDs) can provide a very cost effective solution for short-range illumination. However, achieving good performance at distances of over hundreds of meters requires more directed illumination, such as a focused beam from a laser or specialized spotlight, although special consideration of eye-safety issues is required.

3. Ultraviolet nanostructured detector array development

3.1. Applications of UV imaging technology

Imagery for identification of targets at various distances uses visible cameras, image intensifiers, shortwave IR cameras and long wave uncooled cameras. Each have distinct advantages and disadvantages and are each useful under specific sets of conditions such as light level, thermal conditions, and level of atmospheric obscuration. The shortest wavelength is desired for spatial resolution which allows for small pixels and large formats. [2- 6]

Visible cameras, if adequate light level is present, can provide high resolution, but for long range identification even under moonlit and starlit illuminations, long integration times and large optics are required and dust, smoke and fog easily defeat a single visible camera. Image intensifiers and SWIR cameras are useful in many conditions as the SWIR penetrates fog easily but requires fairly clear night skies for the upper atmospheric airglow light source, and image intensifiers require a certain level of celestial (starlight, moonlight) or light pollution irradiance. Both the SWIR and image intensifiers are limited by the diffraction resolution of the NIR to SWIR wavelengths [5-6].

For optimal resolution, the visible or ultraviolet spectrum is preferable; however, active (laser) illumination is required for long-range night imaging. Covert UV illumination is preferred over the visible and the atmosphere transmits fairly well at the longer UV wavelengths. The covert active system for high-resolution identification modeled in this paper consists of a UV laser source and a silicon CCD, AlGaN or AlGaN APD focal plane array with pixels as small as 4 microns that are spectrally tuned for the solar-blind region of the UV spectrum. The solar-blind region is optimal as virtually all of the solar radiation is absorbed at the higher altitudes leaving a pitch dark terrain even in bright day, yet for sea-level path lengths of 1 km and shorter; the UV atmospheric transmittance is still acceptable.

This combination is ideal for exploitation by a UV illuminator and UV FPA sensor. Current UV lasers can provide either continuous or pulsed energy at levels detectable by solar-blind UV detectors under relatively small optics and at 30 Hz frame rates, providing real-time high-resolution (on the order of 1 cm at 1 km) imagery. At these illumination levels and target ranges, both standard PN, PIN and APD UV detectors and silicon CCD's can be used for target identification. The model has been developed and used to include the combined effects of detector and electronics, atmospheric transmittance and UV background radiance, target size, range and reflectance, and UV laser attributes to simulate and predict both CW and pulsed laser imaging performance and to assist in the design of this prototype system [6].

3.1.1. Model development for passive and active UV systems design

The general equations for SNR prediction for laser illumination and APD are derived.

$$SNR = \frac{Ge_{Lret}}{[F^2G^2(e_{Lret} + e_{bk} + e_{dk}) + (\tilde{e}_{n,amp})^2]^{1/2}} \tag{1}$$

Where G is the APD gain, F is the excess noise, the noise electron terms are the laser return shot noise, the scene noise, the dark current noise and the amp noise

Two special and frequently occurring cases are (2) for the laser power noise limited case and (3) for the amp noise limited case:

$$SNR = \frac{[e_{Lret}]^{1/2}}{F} \tag{2}$$

$$SNR = \frac{Ge_{Lret}}{(\tilde{e}_{n,amp})} \tag{3}$$

The laser return in electrons for cw assuming lambertian reflection is:

$$e_{Lret,ti}=[t_i\eta\tau_o][P_{Lcw}\frac{\lambda}{hc}]\tau_a\frac{\Omega_{pix}}{\Omega_L}\rho_{tar}\tau_a^2\frac{A_o}{\pi R^2}=[t_i\eta\tau_o][P_{Lcw}\frac{\lambda}{hc}]\tau_a\frac{4A_{pix}}{\pi\theta_{fdiv}^2 f^2}\rho_{tar}\tau_a^2\frac{A_o}{\pi R^2} \quad (4)$$

Or when separated into detector/optics, atmosphere, laser and target attributes:

$$e_{Lret,ti}=[t_i\eta\tau_o\frac{A_{pix}A_o}{\pi f^2}]\tau_a^2\frac{[\Phi_{Lcw}]}{\theta_{fdiv}^2}\frac{4\rho_{tar}}{\pi R^2} \quad (5)$$

If we allow for frame summing:

$$e_{Lret,ti+}=N_{fs}[t_i\eta\tau_o\frac{A_{pix}A_o}{\pi f^2}]\tau_a^2\frac{[\Phi_{Lcw}]}{\theta_{fdiv}^2}\frac{4\rho_{tar}}{\pi R^2} \quad (6)$$

For pulse laser operation and using t_{bin} which equals t_{pulse} and the number of bins per frame $N_{bins/f}$:

$$e_{Limage}=[N_{bins/f}\eta\tau_o\frac{A_{pix}A_o}{\pi f^2}]\tau_a^2\frac{P_{Lpulse}(\lambda/hc)}{\theta_{fdiv}^2}\frac{4\rho_{tar}}{\pi R^2} \quad (7)$$

3.1.2. Systems performance metrics for UV systems design

To model the sensor and system performance, we have assumed the pixel size for a high sensitivity, detector size of 5-20 microns for the UV detector array. The fill factor of 70% is assumed typical for these small pixels. Typical quantum efficiencies have been assumed to be in the 70% range for the PIN diode and APD [5-6]. The model uses as default, an amp noise of 15 electrons per frame time, a dark current of 1e-15 amps for a 5 micron pixel or 4 nA/cm^2 or 200 electrons or about 14 noise electrons, and scene noise is effectively zero in the solar-blind region.

The model from the MODTRAN runs shown in figure 3, the daytime irradiance in the UV is insignificant in the solar-blind region. The drop-off from 0.30 microns to 0.26 microns illustrates the requirement for a UV detector with spectral response is in the solar-blind region. Figure 4 shows the UV spectral radiance at midday and the comparative laser illumination of the target at 1 km for a 6 milliradian beam divergence for powers of 1 mW and 10 mW. The left plot in the figure shows that the transmittance improves with longer UV wavelengths for all three levels of aerosols and is sufficient for 1 km lengths in our solar-blind region.

To achieve high-resolution day-night imaging and identification of targets, the following conditions and requirements must be met. While linear detection (no APD and no laser illumination) is fine for muzzle flashes and images of nearby combatants illuminated by live fire (a millisecond event), laser illumination is required for cold targets (facial recognition, profile recognition). A continuous laser and 33 msec integrations are adequate if enough laser power is available. If not, a pulsed laser with nanosecond integrations and APD detectors are required to reduce atmosphere scatter and improve detector sensitivity.

SPECTRAL BANDS		
lam hi	0.267	um
lam lo	0.265	um
lam mid	0.266	um
DETECTOR / FPA		
format	**256**	
dpix	**5**	um
	2.50E-07	cm2
ti cont	**33.33**	msec
t bin	**20**	nsec
t quench	**3000**	nsec
gain	**1**	
gain apd	**1**	
Fm noise	**1**	
amp noise	**5.80E+10**	
tau opt	**0.90**	
qe	**0.70**	
eta inj	**1.00**	
fill factor	**0.75**	
I dark	**5.00E-16**	a
J dark	2.00E-09	a/cm^2
I surface	**1.00E-21**	a
Rload	**1.00E+06**	ohm
resp frame	1.57E-02	
resp bin	9.45E-09	
OPTICS		
dopt	**20**	cm
Aopt	314.16	cm2
focal length	**40**	cm
fnum	2.00	

LASER		
lambda	**0.266**	um
P laser cont	0.10	W
Pd laser cont	3.18E-06	W/cm2
Ph laser cont	1.34E+17	pho/sec
Ph laser cont frame	4.46E+15	pho/frame
E laser pulse	**5.00E-06**	joule/pulse
Ph laser pulse	6.69E+12	pho/pulse
Ph laser bin	6.69E+12	pho/bin
div full	**4**	mrad
t pulse	20	nsec
w pulse	6	meters

SCENE / TARGET TIMING			
t transit	3.33E-06	sec	lin overfill
	3333	nsec	1.25
	3.33	usec	FPA FOV (ft)
N pulses/fr	9999	max poss	5.25

[redacted]

bin and pulse rates	det based	mission based	det/mssion min	min in kHz	used (Hz)	used (kHz)
sample rate (max) Hz	3.31E+05	3.00E+05	3.00E+05	300.0		
N pulses/frame	11036	9999	9999	300.0	666.6	0.6666
N bins/frame	11036	9999	9999	500.0	666.6	0.6666
Dlaser tar (cm, cm^2)	200	3.14E+04				

ELECTRONS AND NOISE from laser, scene, dark current and amp						647.5
	electrons frame full int	electrons for bins in frame	**electrons per bin**	noise e frame full int	noise e for bins in frame	noise e per bin
e laser ret (s)	9713.1	647.5	9.71E-01	98.55	25.45	9.86E-01
e laser ret cont	647.5	*na*	*na*	25.45	*na*	*na*
elect scene (b)	0	9.73E-17	1.46E-19	0.00	0.00	3.82E-10
e dark (d)	104	4.16E-02	6.24E-05	10.20	0.20	7.90E-03
e surface	2.08E-04	8.32E-08	1.25E-10	0.01	0.00	1.12E-05
e kT amp				10.00	0.77	7.74E-03

[redacted]

MODE	**signal**	**noise s+b+d**	**noise b+d**	**SNR s+b+d**	**SNR b+d**	**SNR s+b+d**	**SNR b+d**
DDLM cont	**647.5**	**29.2**	**14.3**	**22.2**	**45.3**	**44.4**	**90.7**
DDLM bins sum	**647.5**	**25.4**	**0.3**	**25.4**	**2267.2**	**50.9**	**4534.3**

Figure 3. UV Sensor Model for evaluating UV Sensor Performance [6]

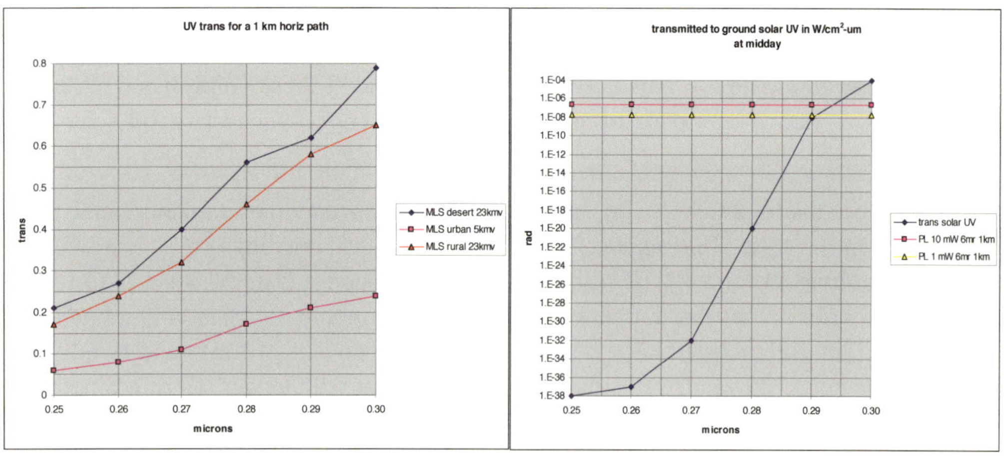

Figure 4. UV transmittance vs. wavelength for three aerosol levels (left) and UV radiance at sea level during midday and laser irradiance on the target at 1 km (6 mradian beam) from a 1mW and 10mW UV laser (right) [6]

http://dx.doi.org/10.5772/58349

3.2. ZnO / MgZnO nanostructures for UV applications

Zinc oxide (ZnO) is a unique wide bandgap biocompatible material system exhibiting both semiconducting and piezoelectric properties that has a diverse group of growth morphologies. Bulk ZnO has a bandgap of 3.37 eV that corresponds to emissions in the ultraviolet (UV) spectral band [7]. Highly ordered vertical arrays of ZnO nanowires (NWs) have been grown on substrates including silicon, SiO_2, GaN, and sapphire using a metal organic chemical vapor deposition (MOCVD) growth process [7]. The structural and optical properties of the grown vertically aligned ZnO NW arrays have been characterized by scanning electron microscopy (SEM), X-ray diffraction (XRD), and photoluminescence (PL) measurements [7-10]. Compared to conventional UV sensors, detectors based on ZnO NWs offer high UV sensitivity and low visible sensitivity, and are expected to exhibit low noise, high quantum efficiency, extended lifetimes, and have low power requirements [11-12]. The Photoresponse switching properties of NW array based sensing devices have been measured with intermittent exposure to UV radiation, where the devices were found to switch between low and high conductivity states at time intervals on the order of a few seconds. Envisioned applications for such sensors/FPAs potentially include defense and commercial applications [13].

Zinc oxide is a versatile functional material that provides a biocompatible material system with a unique wide direct energy band gap and exhibits both semiconducting and piezoelectric properties. ZnO is transparent to visible light and can be made highly conductive by doping. Bulk ZnO has a bandgap of 3.37 eV that includes emissions in the solar blind ultraviolet (UV) spectral band (~240-280 nm), making it suitable for UV detector applications [7]. Over this wavelength range, solar radiation is completely absorbed by the ozone layer of the earth's atmosphere, so the background solar radiation at the earth's surface is essentially zero. This enhances the capability of UV sensors in missile warning systems to detect targets such as missile plumes and flames emitting in this region.

ZnO is the basis for the one of the richest families of nanostructures among all materials taking into accounts both structure and properties. ZnO growth morphologies have been demonstrated for nanowires, nanobelts, nanocages, nanocombs, nanosprings, nanorings, and nanohelixes [7]. The development of ZnO nanowire (NW) based UV detectors offers high UV sensitivity and low visible sensitivity for missile warning related applications. Demonstration of devices using single ZnO NW strands has been widely reported in literature [7-16]. However, the development of reliable 2D arrays of aligned ZnO NWs has proven more challenging. The demonstration of reliable 2D arrays requires (1) correlation of growth process and growth parameters with the material quality of ZnO NWs, (2) correlation of the electrical and optical performance with growth parameters and fabrication processes, and (3) addressing system design challenges [17-18].

With conventional NW growth methods including electrochemical deposition, hydrothermal synthesis, and molecular beam epitaxy (MBE), it is generally difficult to scale up and control NW growth. Electrochemical deposition is well suited for large scale production but does not allow control over the NW orientation. Hydrothermal synthesis is a low temperature and low-cost process that allows growth of NWs on flexible substrates without metal catalysts, but the direction and morphology of the NWs cannot be well-controlled with this method [8-10]. The

MBE method allows monitoring of the structural quality during NW growth; however, this type of synthesis often requires use of metal catalysts as a seed layer [10], which introduces undesired defects to the structure, decreasing the crystal quality [12-16]. Chemical vapor deposition (CVD) also requires catalysts at the NW tips, and using this method the tips of the grown NWs were observed to be flat, with vertical alignment.

3.3. Characterization of ZnO NWs arrays grown on the various substrates

The samples were characterized by scanning electron microscopy (SEM) utilizing a Quanta FEG 250 system, and X-ray diffraction (XRD) using Bruker D-8 Advance X-ray diffractometer with a wavelength of 1.5406 Å corresponding to the Cu Kα line. In addition, photoluminescence (PL) measurements were performed at room temperature using a Linconix HeCd UV laser emitting at a wavelength of 325 nm. A Si detector in conjunction with at lock-in amplifier and chopper were used to measure the PL from the beam reflected off the sample at the output over the desired wavelength range [18-20].

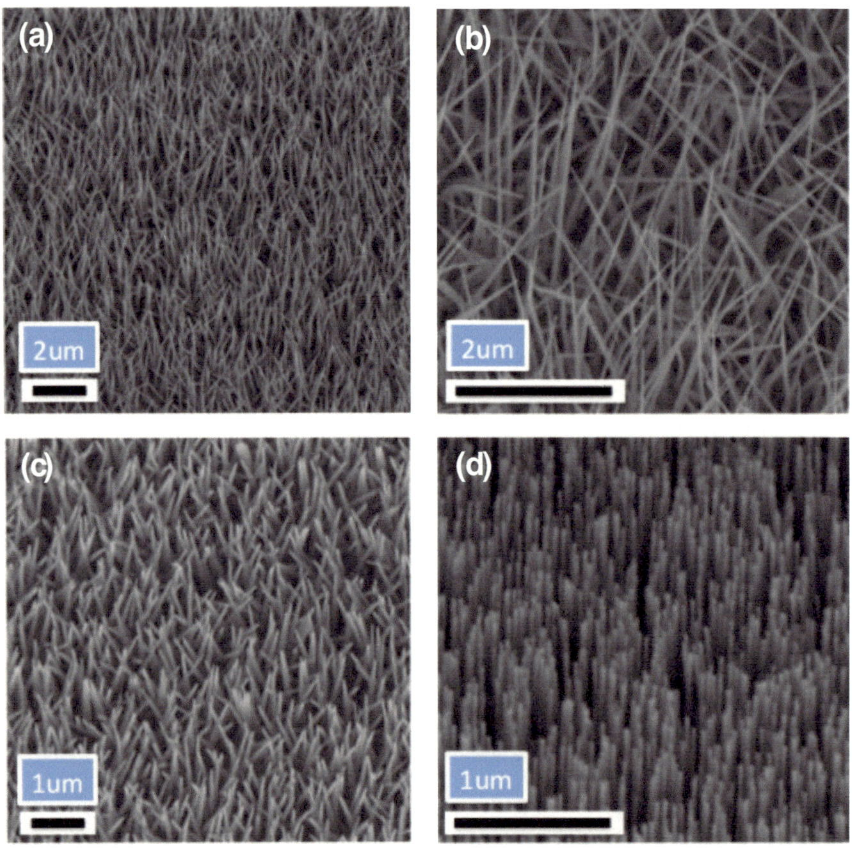

Figure 5. Scanning electron microscope (SEM) images of NWs grown on the various substrates taken at room temperature, showing NWs grown on (a) ZnO/sapphire; (b) ZnO/SiO_2/p-Si; (c) ZnO/p-Si; and (d) ZnO/GaN/sapphire.[20]

SEM was performed to explore the NWs morphology. Figure 5 show the synthesized ZnO NWs on the various substrates, which can be generally seen to have uniform distribution density. The ZnO NWs grown on sapphire [Figure 5(a)] had approximate diameters of 50-70 nm and lengths in the range of 1-2 μm. NWs grown on SiO_2 [Figure 5(b)] had diameters of 150-200 nm and lengths of 1-2 μm, and were the least vertically oriented and associated with a relatively high lattice mismatch. NWs grown on the Si (111) substrate [Figure 5(c)] had a slightly random orientation, also having diameters in the range of 150-200 nm and lengths from 1-2 μm. Finally, the NWs grown on GaN [Figure 5(d)] showed strong vertical orientation, with diameters of 20-40 nm and lengths of 0.7-1.0 μm [20].

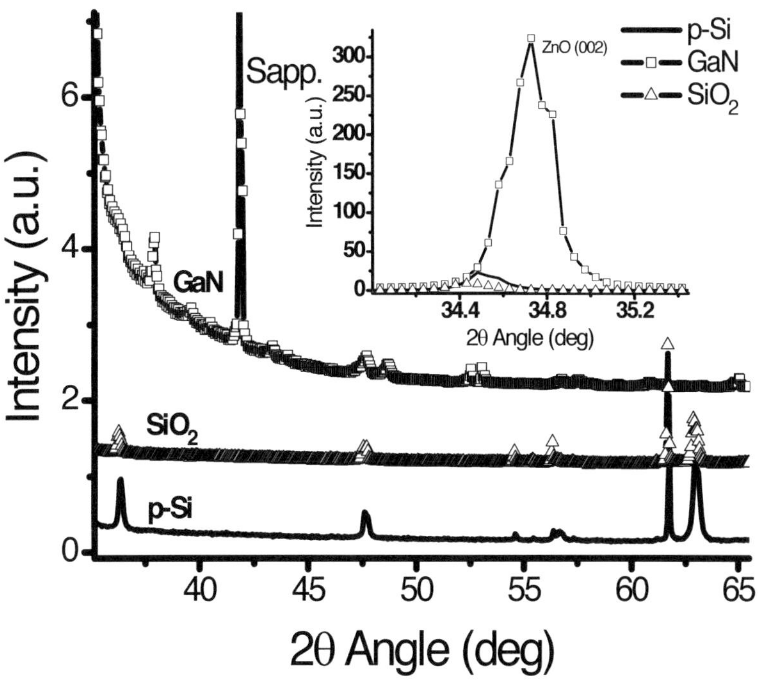

Figure 6. X-ray diffraction (XRD) of ZnO NWs grown using MOCVD on p-Si (solid), GaN/sapphire (square) and SiO_2 (triangle). The inset shows the ZnO peak associated with ZnO oriented along (002) and GaN [20].

Figure 6 shows the XRD pattern for the ZnO NWs grown on p-Si, GaN, and SiO_2 substrates [10]. The inset of Figure 2 shows dominant peaks related to ZnO (002). The peak at 34º (2θ) for ZnO grown on p-Si and SiO_2 substrates incorporated the overlapping of ZnO NWs (002) and

ZnO thin film (002). An additional diffraction peak associated with GaN was present for the GaN/sapphire substrate. ZnO NWs oriented along the (002) direction had full-widths at half maxima (FWHM) and c-lattice constants of 0.0498 (θ) and 5.1982 Å at 34.48° (2θ) for p-Si, 0.0497(θ) and 5.1838 Å at 34.58° (2θ) for GaN, 0.0865(θ) and 5.1624° at 34.38º (2θ) for SiO_2, and 0.0830˚(θ) and 5.2011 Å at 34.46º (2θ) for sapphire.

The quality of the ZnO epilayers utilized as seed layers to grow ZnO NWs was also characterized. The ZnO thin films were oriented along (002) and had a maximum at 34.58º with FWHM of 0.0697 (θ) for p-Si, maximum of 34.58º with FWHM of 0.0684 (θ) for GaN, and maximum of 34.43º with FWHM of 0.0557 (θ) for SiO_2. Additional shallow diffraction peaks were observed for NWs grown on p-Si and SiO_2, which are attributed to ZnO (100, 101, 102 and 110) as can be seen from Figure 6. As shown in Figure 7, for ZnO NW growth on sapphire major peaks were observed for ZnO (002) at 34.46° (2θ) and Al_2O at 37.91° (2θ), with a minor peak for ZnO (101) at 36.34° (2θ).

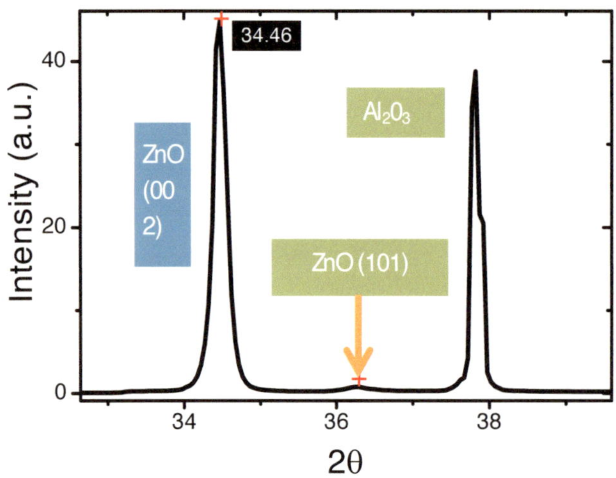

Figure 7. XRD of ZnO NWs grown using MOCVD on sapphire [20]

3.4. Photoluminescence (PL) measurements

Figure 8 shows the PL spectra for ZnO NWs grown on p-Si, GaN, and SiO_2 substrates [10]. The room temperature PL measurements were performed using a ~280 nm light source. Single peaks located at 380 nm having a FWHM of 14.69 nm and at 378 nm having a FWHM of 15 nm were observed for p-Si and SiO_2 substrates, respectively, corresponding to the recombination of excitons through an exciton-exciton collision process [18-20].

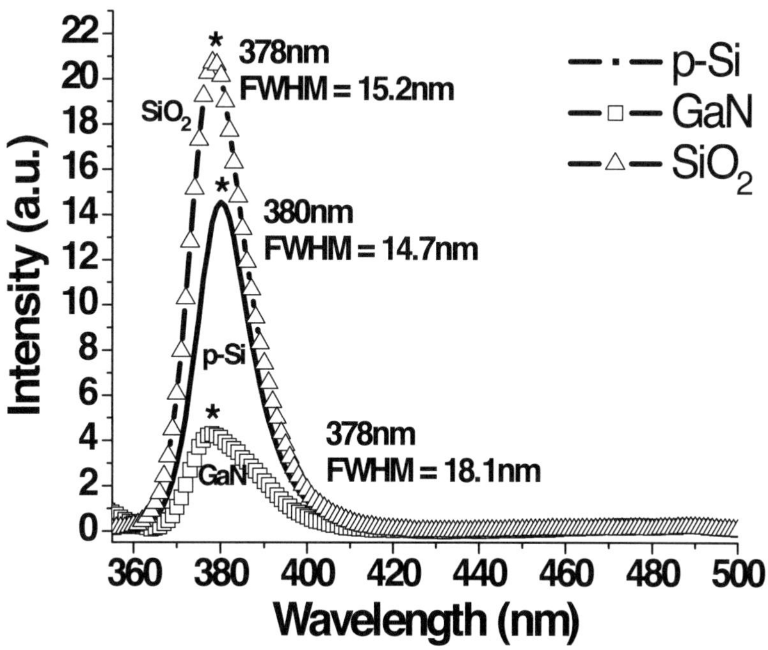

Figure 8. Photoluminescence (PL) of ZnO NWs grown on p-Si (100) (solid) with a single peak at 380 nm, GaN (square) with a stronger peak at 378 and SiO_2 (triangle) with a single peak at 378 nm [20].

No defects related to Zn or O vacancies were observed, which can be attributed to the confinement of defects at the ZnO thin film/substrate interface. For the ZnO NWs grown on GaN, a predominant peak with a FWHM of 18.18 nm was observed at 378 nm. High stress was evident for ZnO NWs grown on GaN, which can be observed in Figure 2; this can contribute to the broadening of the peak in comparison to p-Si and SiO_2. Shallow peaks identified at 474 nm and 490 nm through Lorentzian decomposition are attributed to oxygen interstitial and oxygen vacancies, respectively [20].

A UV LED lamp acquired from Sensor Electronic Technology Inc. was used to characterize the UV Photoresponse of the ZnO NW arrays [20]. The lamp comprises eight separate AlGaN based UV LEDs in a TO-3 package spanning the 240-370 nm wavelength range, with a customized power supply capable of independently monitoring and controlling the current of all or any of the LEDs. The Photoresponse was determined by first applying voltage between indium contacts on the front and back sides of a Si NW sample and measuring the resulting current in the dark, and then repeating this procedure while the sample was exposed to radiation from a UV LED at a specific wavelength.

Figure 9 shows the on-off switching characteristics of a ZnO vertical array NW device when exposed to radiation at 370 nm. This device was found to switch between low and high

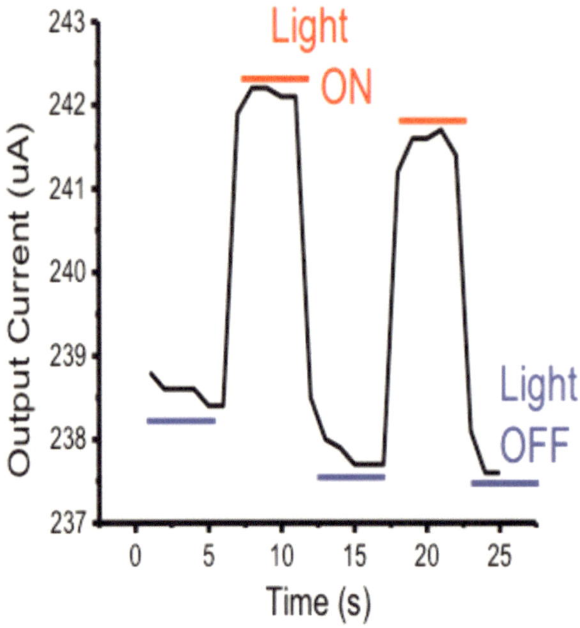

Figure 9. Switching Photoresponse characteristics of ZnO NW device when UV LED source at ~370 nm turned on and off over approximately 10 s intervals.[20]

conductivity states in approximately 3 s, a faster response than most reported thus far for ZnO NW based UV detectors.

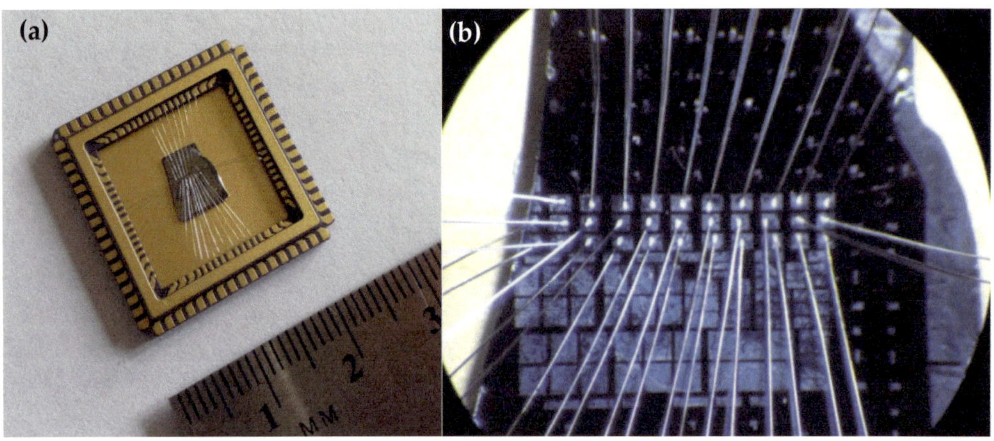

Figure 10. (a) Mounted solar blind NW UV 3x9 pixel array detector device; (b) close-up of device, showing wire bonded pixels [20].

Figure 9(a) shows a mounted and wire bonded NW UV 3x9 pixel array detector device. Incorporation of Mg allows the detector response to be shifted to shorter wavelengths to provide detection in the solar blind region. This device was tested by applying a bias between the top contacts on the pixels, which are apparent in Figure 9(b), and the back contact.

ZnO nanowires based arrays offer high sensitivity and have potential application in UV imaging systems. ZnO nanowire array based UV detectors have no moving parts, high quantum efficiency, extended lifetimes, low noise, low power requirements, and offer high sensitivity.

ZnO nanowires have also been evaluated for providing remote power for the stand alone sensors. This type of application has been extensively studied by Professor Z.L. Wang and his team at Georgia Tech [21, 22]. They have shown that ZnO nanowires can be used as nano-generators for providing remote power using the Piezo-electric effect. Photovoltaic cells or solar cells are a popular renewable energy technology, relying on approaches such as inorganic p-n junctions, organic thin films, and organic-inorganic heterojunction. However, a solar cell works only under sufficient light illumination, which depends on the location the devices will be deployed, as well as the time of the day and the weather.

Considering that mechanical energy is widely available in our living environment, They have demonstrated [21] the first hybrid cell for concurrently harvesting solar and mechanical energy through simply integrating a dye-sensitized solar cell (DSSC) and a piezoelectric nanogenerator on the two sides of a common substrate. After this, in order to solve the encapsulation problem from liquid electrolyte leakage in the first back-to-back integrated HC, early in 2011, Xu and Wang improved the prototype design of the HC and developed a compact solid state solar cell. This innovative design convoluted the roles played by the NW array to simultaneously perform their functionality in a nanogenerator and a DSSC. The design and the performance are shown in figure 11.

Based on these demonstrations of HCs for concurrently harvesting solar and mechanical energies, they have. reported an optical fiber-based three-dimensional (3D) hybrid cell, consisting of a dye-sensitized solar cell for harvesting solar energy and a nanogenerator for harvesting mechanical energy; these are fabricated coaxially around a single fiber as a core–shell structure (Figure 11). The optical fiber, which is flexible and allows remote transmission of light, serves as the substrate for the 3D DSSC for enhancing the electron transport property and the surface area, and making it suitable for solar power generation at remote/concealed locations. The inner layer of the HC is the DSSC portion, which is based on a radically grown ZnO NW array on an optical fiber with ITO as the bottom electrode. The dye-sensitized ZnO NW array was encapsulated by a stainless steel capillary tube with a Pt-coated inner wall as the photo- anode for the DSSC. The stainless steel tube also serves as the bottom electrode for the outer layer of the nanogenerator, with densely packed ZnO NWs grown on its outer wall.

Another exciting application of ZnO nanowires is designing, fabricating, and integrating arrays of nanodevices into a functional system are key to transferring nanoscale science into applicable nanotechnology as shown in Figure 12.

Recent work [22] on three-dimensional (3D) circuitry integration of piezotronic transistors based on vertical zinc oxide nanowires as an active taxel-addressable pressure/force sensor matrix for tactile imaging. Using the piezoelectric polarization charges created at a metal-semiconductor interface under strain to gate/modulate the transport process of local charge carriers, we designed independently addressable two-terminal transistor arrays, which convert mechanical stimuli applied to the devices into local electronic controlling signals.

The device matrix can achieve shape-adaptive high-resolution tactile imaging and self-powered, multidimensional active sensing. The 3D piezotronic transistor array may have applications in human-electronics interfacing, smart skin, and micro- and nano-electromechanical systems.

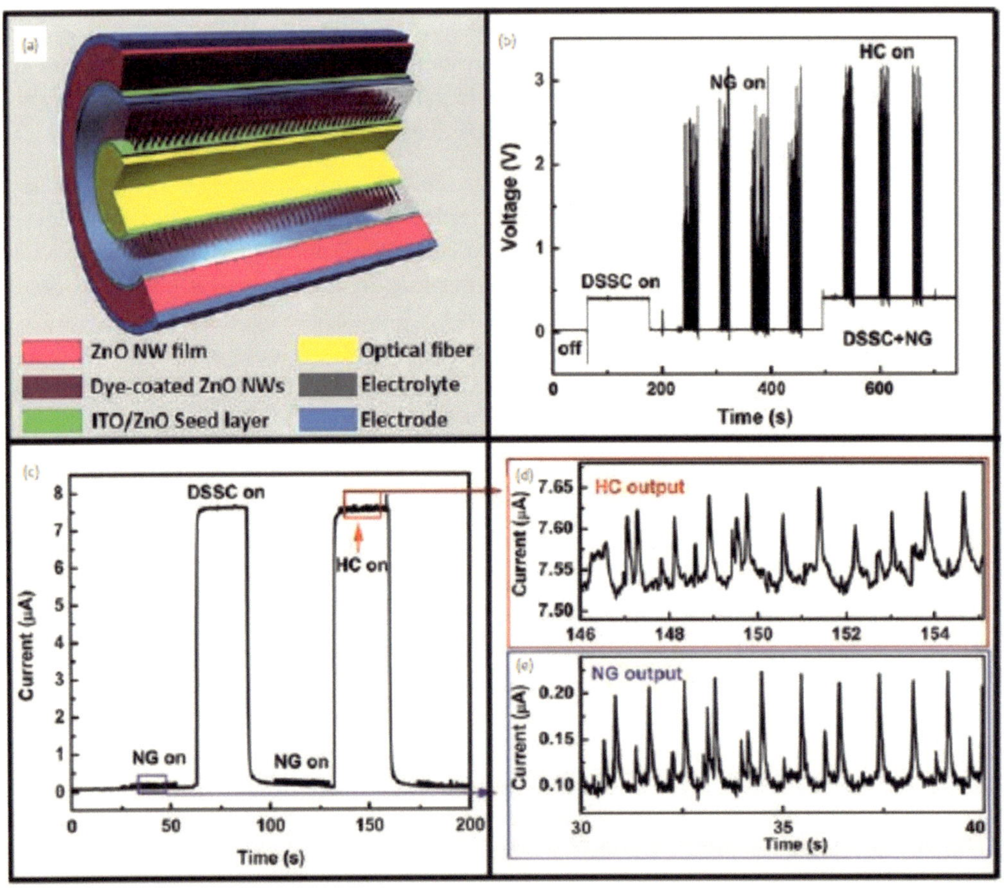

Figure 11. Design and performance of a 3D optical fiber based hybrid cell (HC) consisting of a dye-sensitized solar cell (DSSC) and a nanogenerator (NG) for harvesting solar and mechanical energy. (a) The 3D HC is composed of an optical fiber based DSSC with capillary tube as counter electrode and a NG on top. (b) Open-circuit voltage (VOC) of the HC when the NG and the DSSC are connected in series, where VOC(HC)= VOC(DSSC)+ VOC(NG). (c) Short-circuit current (ISC) of the HC when the NG and the DSSC are connected in parallel. (d) and (e) Enlarged view of ISC(HC) and ISC(NG), clearly showing that ISC(NG) is 0.13 μA, the ISC(DSSC) is 7.52 μA, and the ISC(HC) is about 7.65 μA, nearly the sum of the output of the solar cell. [21].

http://dx.doi.org/10.5772/58349

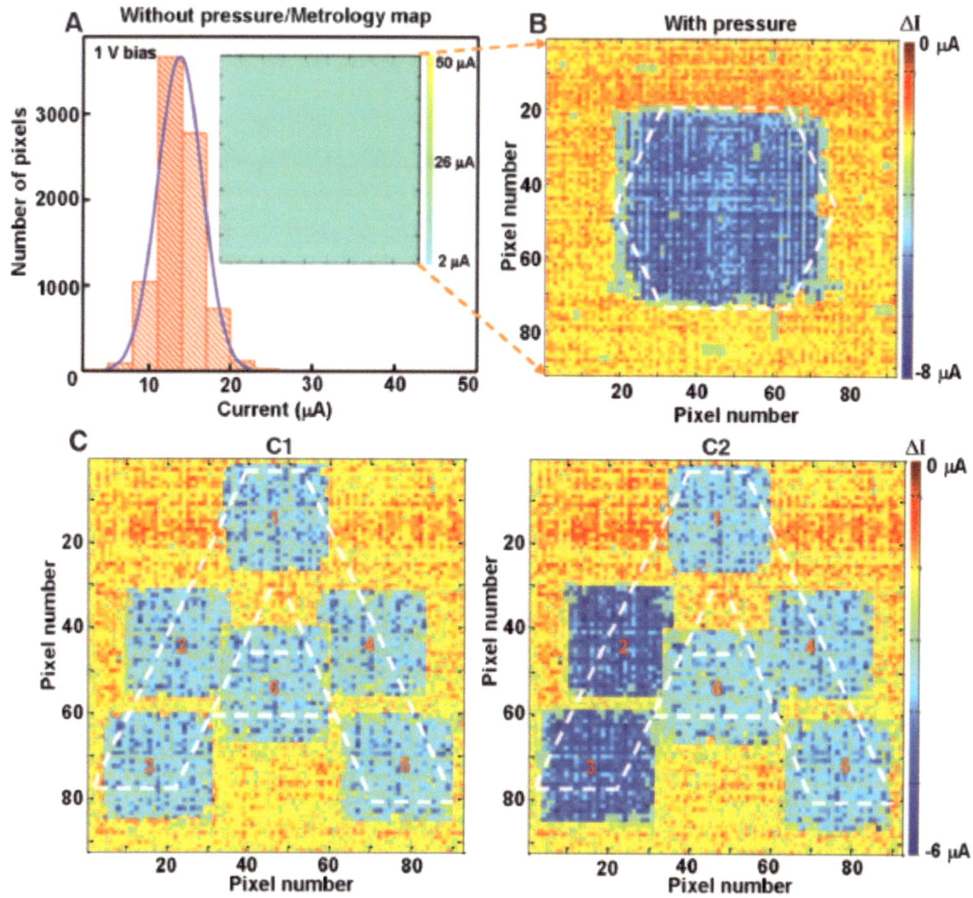

Figure 12. Tactile imaging and multidimensional sensing by the fully integrated 92 × 92 SGVPT array. (A) Metrology mapping (inset) and statistical investigation of the fully integrated SGVPT array without applying stress. (B) Current response contour plot illustrating the capability of SGVPT array for imaging the spatial profile of applied stress. Color scale represents the current differences for each taxel before and after applying the normal stress. The physical shape of the applied stress is highlighted by the white dashed lines. (C) Multidimensional sensing by an SGVPT array exhibits the potential of realizing applications such as personal signature recognition with maximum security and unique identity. The shape of a "written" letter A is highlighted by the white dashed lines. [22].

4. Development of GaN UVAPD for ultraviolet sensor applications

High resolution imaging in UV bans has a lot of applications in Defense and Commercial applications. The shortest wavelength is desired for spatial resolution which allows for small pixels and large formats. UVAPD's have been demonstrated as discrete devices demonstrating gain. The next frontier is to develop UV APD arrays with high gain to demonstrate high resolution imaging. We will discuss model that can predict sensor performance in the UV band using APD's with various gain and other parameters for a desired UV band of interest. SNR's can be modeled from illuminated targets at various distances with high resolution under standard atmospheres in the UV band and the solar blind region using detector arrays with unity gain and with high gain APD's [23-26].

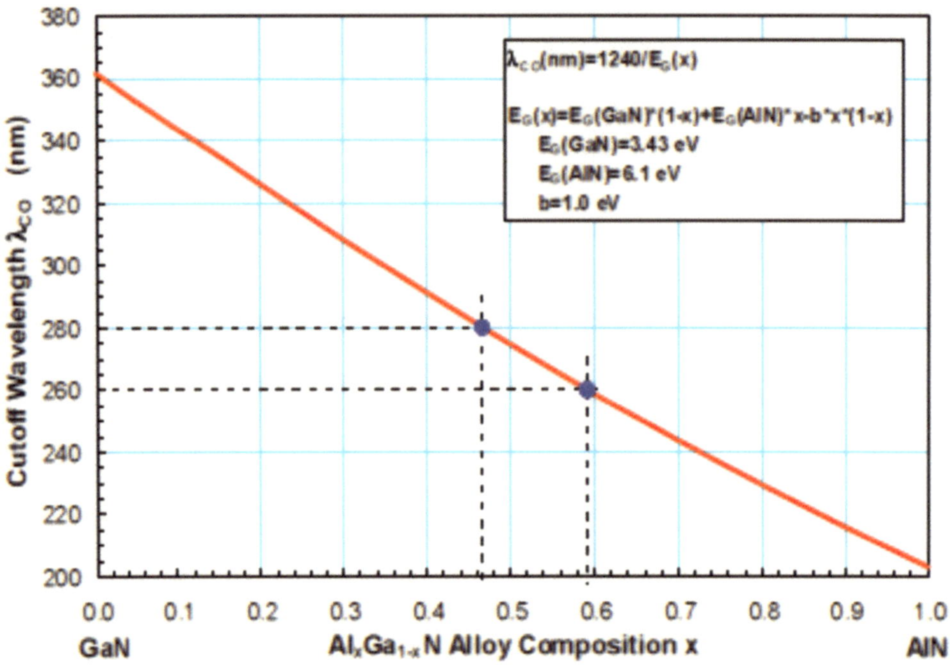

Figure 13. Relationship between alloy composition of AlGaN and the corresponding spectral cutoff for the UV detector arrays [23].

Figure 13 presents the relationship between the alloy composition of Gallium and Aluminum in $Al_xGa_{1-x}N$ that determines the cut-off wavelength of the UV detector for p-i-n [23-24] and also for UV APD's. Deep Ultra Violet (DUV) will require addition of larger composition of Aluminum in $Al_xGa_{1-x}N$. [25].

5. GaN /AlGaN UV APD growth

Figure 14 presents the High-Temperature MOCVD system by Aixtron. This new reactor design and capability has the ability to grow high quality GaN and AlGaN material with doping for GaN/AlGaN UVAPD applications [26].

Figure 15 presents the device structure of a back-side illuminated AlGaN UV APD. The substrate in this device structure is double side polished AlN substrate. The use of AlN substrate allows the UV APD device structure to be back-side illuminated and can be integrated with silicon CMOS electronics. Figure 16 presents the Reciprocal Space mapping of AlGaN on AlN substrate and Sapphire substrate. The data for sapphire substrate shows increased strain and mosaicity compared with AlN substrate.

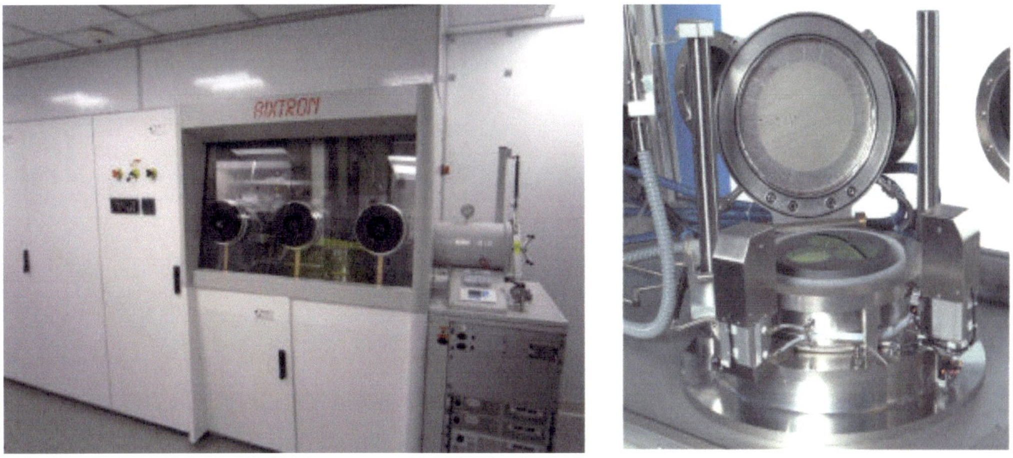

Figure 14. Photograph of New-generation AIXTRON CCS 3x2 " high temperature III-Nitride 3x2 MOCVD growth Chamber open for loading wafers showing close-coupled showerhead [26]

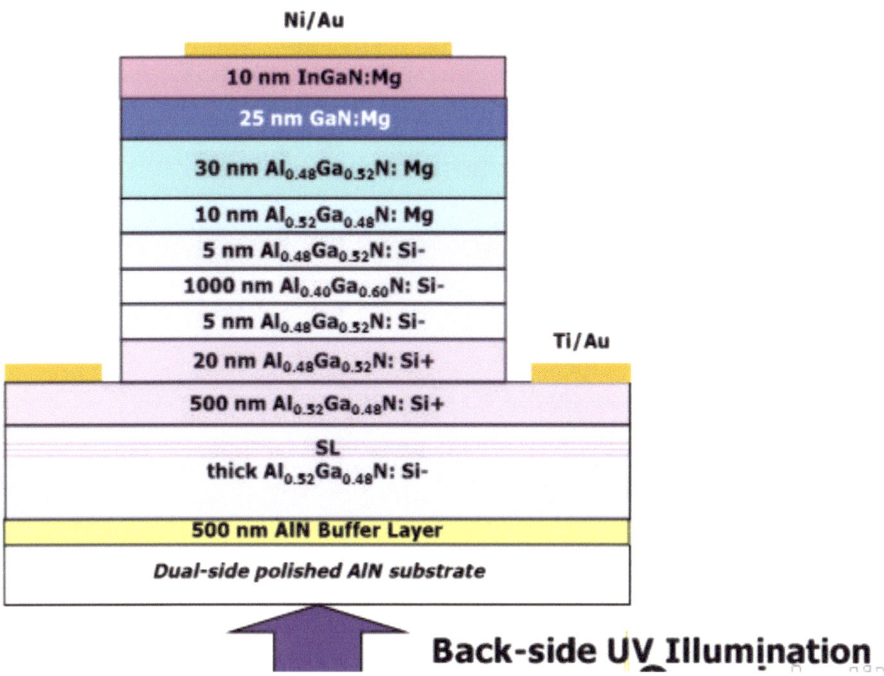

Figure 15. Device Structure Cross-section of prototype Back-Side Illuminated AlGaN UV APD [27]

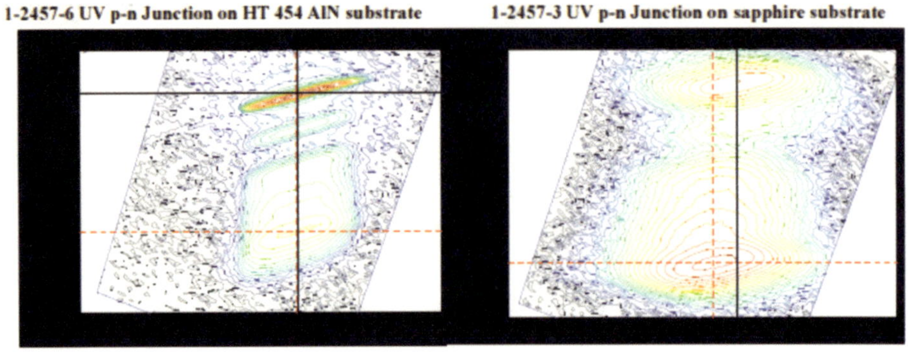

Figure 16. Reciprocal Space mapping of AlGaN p-n junctions on AlN and Sapphire Substrates [28]

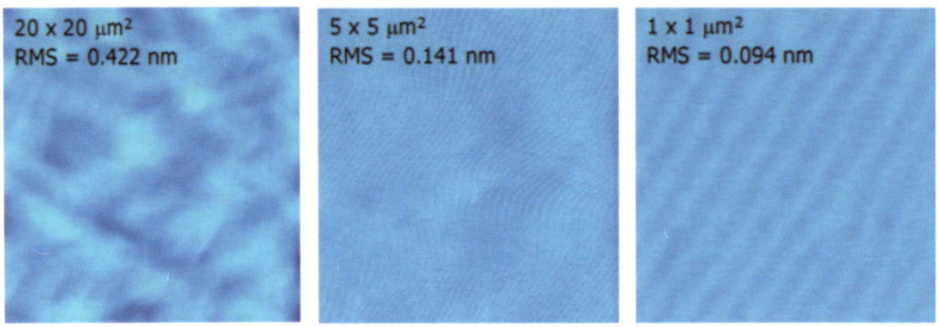

Figure 17. Microscopic surface morphology using AFM on GaN p-i-n structure grown on GaN/Sapphire template. No specific surface defects are observed [27]

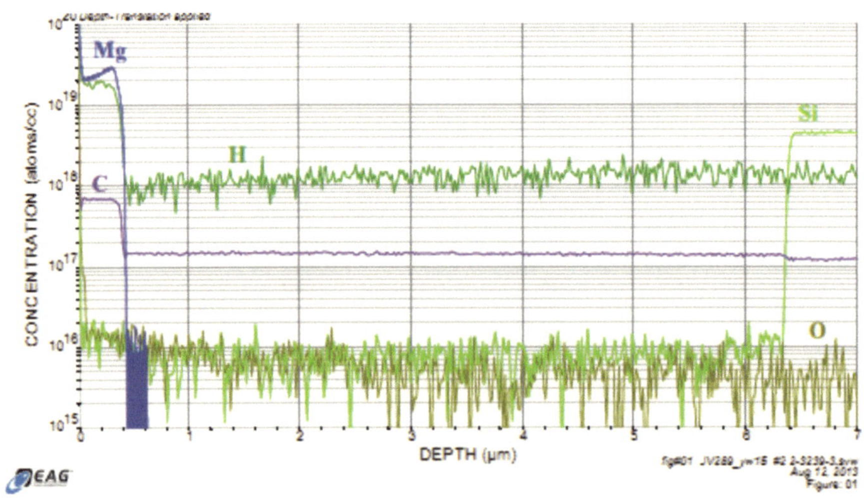

Figure 18. SIMS analysis of GaN p-i-n structure on GaN/Sapphire template, the data shows controlled Si and Mg doping for n- and p-type layers. The data shows low background doping concentration in GaN layer [28]

http://dx.doi.org/10.5772/58349

Figure 17 presents the microscopic surface morphology using AFM on GaN p-i-n structure grown on GaN/Sapphire template. No surface defects are observed. These results are encouraging to develop a low cost back-side illuminated UV APD detector array.

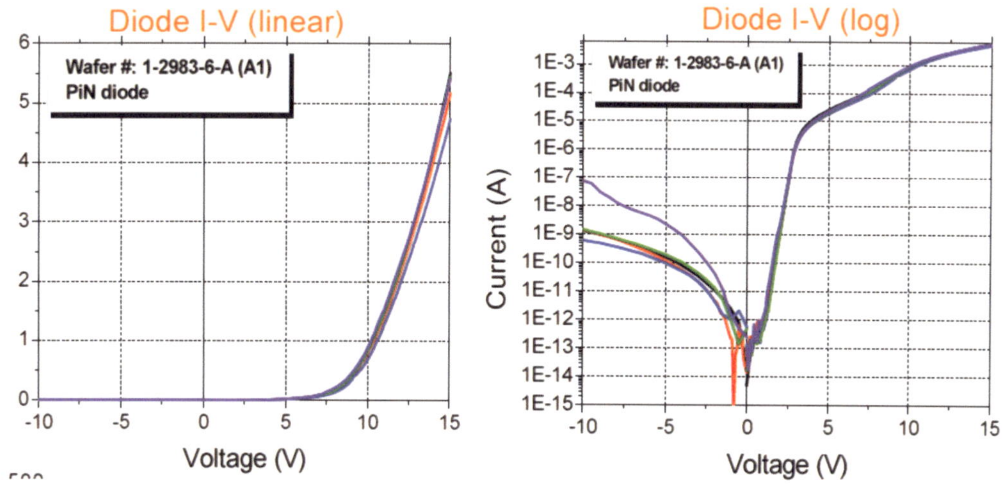

Figure 19. Current - Voltage characteristics of AlGaN UV APD Unpassivated test structure. Further reduction in the dark current will be expected with surface passivation [29].

Figure 18 presents the SIMS analysis of GaN p-i-n structure on GaN/Sapphire template, the data shows controlled Si and Mg doping for n- and p-type layers. The data shows low background doping concentration in GaN layer. The Mg doping is being increased for better p-type conductivity

Figure 19 presents the Current-Voltage characteristics of AlGaN UV APD with spectral response of 300 nm. Further reduction in the dark current will be expected with surface passivation. The future effort is underway to improve the growth characteristics Low-defect-density in substrates and high-quality epitaxial growth technologies are the keys to the successful implementation of a robust high-performance APDs for UV focal plane arrays [29-33].

6. Visible –Near Infrared (NIR) detector technologies

The Visible-Near Infrared band (0.4-1.7 μm) bridges the spectral gap between the visible and thermal bands in the electromagnetic spectrum. In this spectral band, the primary phenomenology of interest is the reflectance signature of the target, manifested as either its variations in brightness or spectral reflectance, or both.

Infrared imaging in the NIR /SWIR band offers several advantages: it can detect reflected light, offering more intuitive, visible-like images; is better suited for imaging in adverse environments and weather conditions, including fog, dust, and smoke; can also see in low light conditions, and use eye safe 1550 nm illumination; and can generate digital video outputs and thus offering a more dynamic range than traditional image intensifier night vision equipment. Under low light conditions, the sensitivity of the focal plane array is ultimately determined by the R_0A product of the photodiode.[34-36].

6.1. Si_{1-x} Ge_x (SiGe) detector arrays

Like the other two alloy semiconductors mentioned above, SiGe is another example of material that can be used for the fabrication of IR detectors. The key attractive feature of SiGe IR detectors is that they can be fabricated on large diameter Si substrates with size as large as 12-inch diameter using standard integrated circuit processing techniques. Furthermore, the SiGe detectors can be directly integrated onto low noise Si ROICs to yield low cost and highly uniform IR FPAs.

Some of the earlier attempts in developing SiGe IR detectors focused on their LWIR applications [34-36]. Renewed efforts are now developing these detectors for application in the NIR-SWIR band [36]. For the SiGe material to respond to the SWIR band, its cutoff wavelength is tuned by adjusting the SiGe alloy composition. Si and Ge have the same crystallographic structure and both materials can be alloyed with various Ge concentration. The lattice constant of Ge is 4.18% larger than that of Si, and for a Si_{1-x} Ge_x alloy the lattice constant does not exactly follow Vegard's law. The relative change of the lattice constant is given by [36]:

$$a_{Si1-x\ Gex} = 0.5431 + 0.01992x + 0.0002733x^2 (nm).$$

For a Si_{1-x} Ge_x layer with $x > 0$ on a Si substrate means that the layer is under compressive stress. A perfect epitaxial growth of such a strained heteroepitaxial layer can be achieved as long as its thickness does not exceed a critical thickness for stability. Beyond the critical thickness, the strain is relaxed through the formation of misfit dislocations which can cause an increase in the dark current.

Several approaches have been proposed to reduce the dark current in SiGe detector arrays by several orders of magnitude; these include Superlattice, Quantum dot and Buried junction designs [36-38]. Furthermore, some of these approaches have the potential of extending the wavelength of operation beyond 1.8-2.0 microns. The challenge is to take advantage of these innovative device designs and reduce the dark currents to 1-10 nA cm^{-2}. Figure 20 presents the SiGe /Ge detector array using buried junction approach to reduce the surface states and leakage current [36].

Figure 22 shows the Strained-Layer Superlattice (SLS) structure being evaluated for longer detector array response to 2 microns.. It consists of SiGe quantum wells and Si barrier layers, grown on p-type (001) Si substrates. Super lattices having differing Si barrier and Ge well thicknesses to control the strain are grown to optimize wavelength response and dark current.

The SiGe well thicknesses are kept below the critical layer thickness for dislocation formation. To complete the structure, the undoped superlattice is capped with a thin n+ Si cap layer to form the p-n junction. After growth the devices are patterned with a top contact, mesas are etched to provide isolation and the substrate contact is formed. The etched mesa can also be passivated to minimize surface recombination as indicated in Figure 22. The device shown in the figure 22 uses substrate illumination, as is needed for use in FPA arrays, and short wavelength response can be improved by thinning the Si substrate.

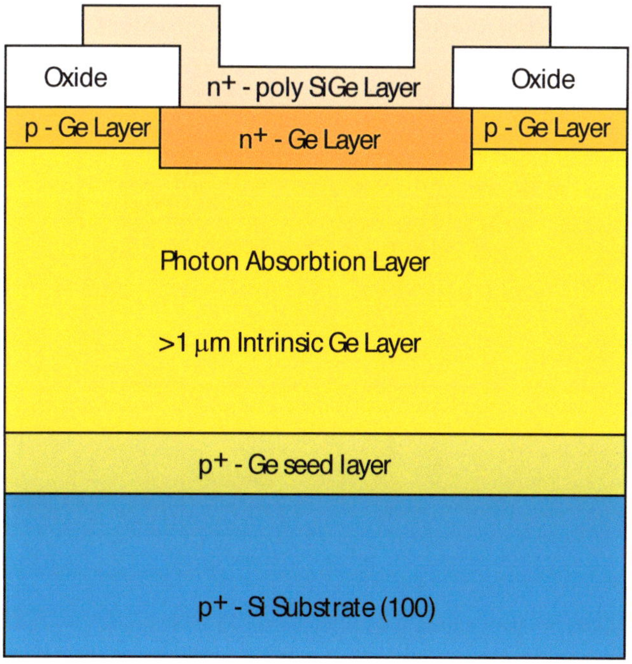

Figure 20. SiGe/Si based buried junction approach to be evaluated for reduced surface states and leakage current [36]

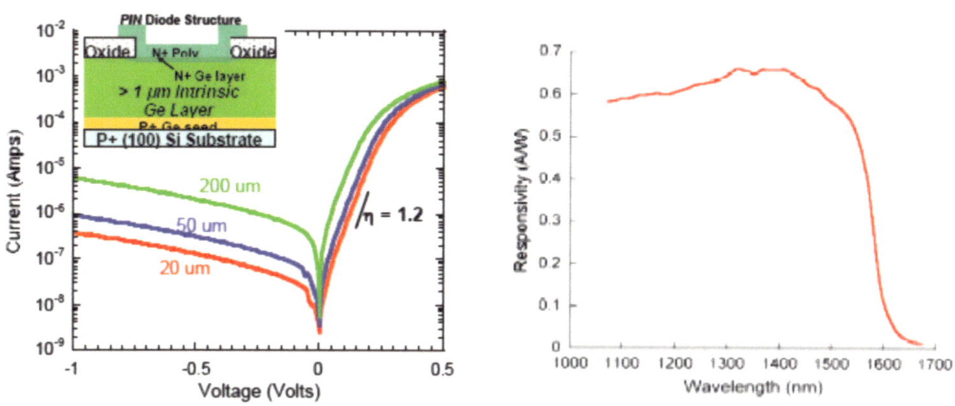

Figure 21. Measured room temperature I-V characteristics for large area diodes with 20, 50 and 200 micron unit cell. The inset shows the schematic device cross section. The spectral response data for SiGe detector is also presented. [38]

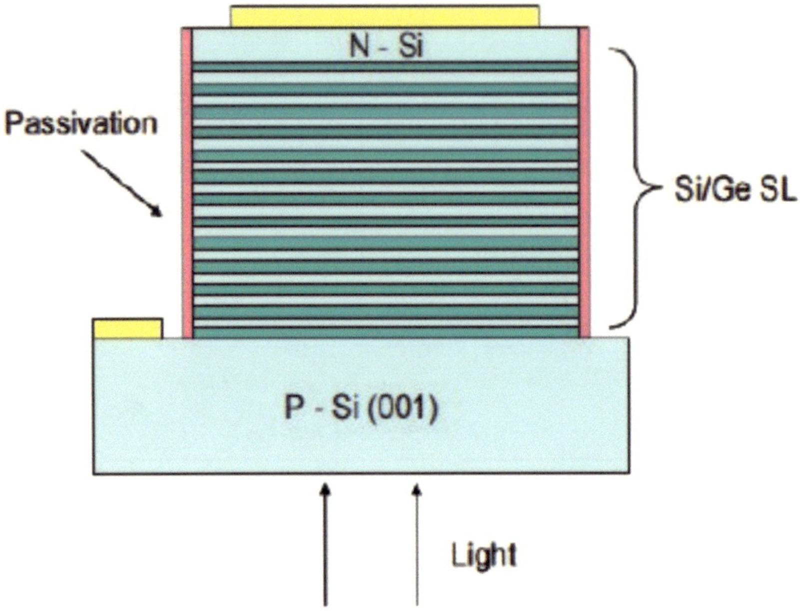

Figure 22. Schematic of detector array structure consisting of a SiGe /Si strained layer Superlattice grown on (001) silicon [38].

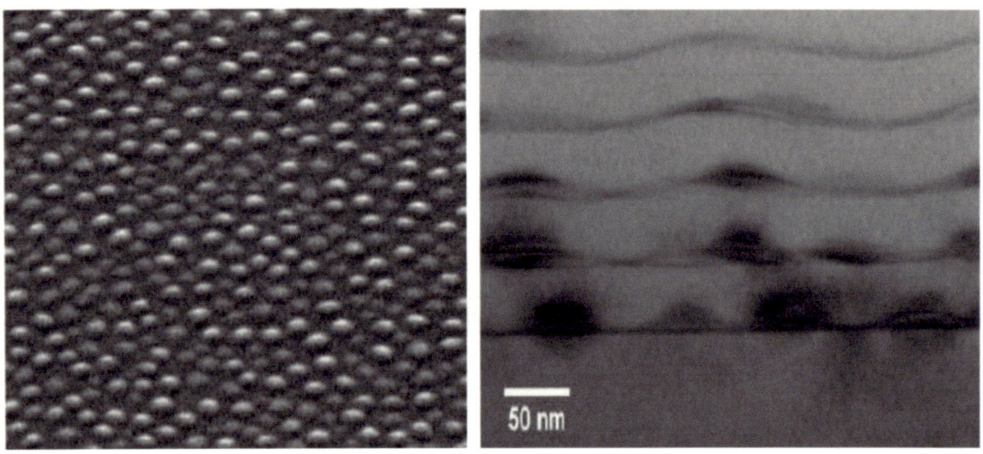

Figure 23. SEM image (45° tilt) of a Ge QD layer deposited on Si. The QDs are ~60 nm in diameter with a density of 1020 cm2. Also shown is a Cross-sectional TEM image of Ge/Si QDSL grown. Ge QDs appear with dark contrast compared to Si barriers. [38].

http://dx.doi.org/10.5772/58349

The strained-layer superlattice and quantum dot superlattice (QDSL) in the SiGe material system have the potential of developing Vis-NIR detector arrays with longer cutoff wavelength and potentially lower dark current. The advantage of quantum dots is the potential to exploit the optical properties of Ge while avoiding dislocation formation. Ge QDs grown on Si in Stranski-Krastanov mode can be deposited well beyond the critical thickness without dislocation nucleation [39].

Figure 23 shows an SEM image of an array of Ge nanodots grown by MOCVD. These dots are typically 50-75 nm in diameter with area coverage of ~20%. To increase optical absorption and sensitivity, MOCVD-based growth techniques is being developed for the deposition of Ge/Si quantum dot superlattices (QDSLs), where Ge QDs are alternated with thin (10-30 nm) Si barrier layers. A cross-sectional TEM image of QDSLs is shown in Figure 23.b.

7. SWIR detector technologies

The SWIR band (0.9-2.5 μm) bridges the spectral gap between the visible and thermal bands in the electromagnetic spectrum. In this spectral band, the primary phenomenology of interest is the reflectance signature of the target, manifested as either its variations in brightness or spectral reflectance, or both.

Infrared imaging in the SWIR band offers several advantages: it can detect reflected light, offering more intuitive, visible-like images; is better suited for imaging in adverse environments and weather conditions, including fog, dust, and smoke; can also see in low light conditions, and use eye safe 1550 nm illumination that is totally undetectable by regular night vision equipment; and can generate digital video outputs and thus offering a more dynamic range than traditional image intensifier night vision equipment. Under low light conditions, the sensitivity of the focal plane array is ultimately determined by the R_0A product of the photodiode.

7.1. $In_x Ga_{1-x}As$ detector array development

For SWIR imaging, InGaAs is one of the widely used detector materials due to its low dark current. The detector material can be prepared using any of the following techniques: Molecular beam epitaxy (MBE), metal-organic chemical vapor deposition (MOCVD), liquid phase epitaxy (LPE), hydride-transport vapor phase epitaxy (VPE), and atomic layer epitaxy (ALE). InGaAs layers are typically grown on lattice matched InP substrates using an alloy composition of x = 0.53 [40-42].

The spectral response typically covers 0.9-1.7μm at room temperature. By increasing the composition to x=0.82, InGaAs is able to extend its cutoff to 2.6 μm. However, the crystal defects due to epitaxy and the decreased shunt resistance, due to a smaller band gap, degrade performance at the longer cutoff wavelengths. [43].

The band gap [7-1] of the strained $In_x Ga_{1-x} As$: InP structure can be tailored by varying the alloy composition during crystal growth according to the equation:

$$E_g(eV) = \left(E_{g_{GaAs}} - \frac{\alpha_{GaAs}T^2}{T + \beta_{GaAs}} + \left(E_{g_{InAs}} - \frac{\alpha_{InAs}T^2}{T + \beta_{InAs}} - E_{g_{GaAs}} + \frac{\alpha_{GaAs}T^2}{T + \beta_{GaAs}}\right)x - 0.475x(1 - x)\right)$$

Where E_g is the band gap in (eV), α and β are fitting parameters, and x is the In: As ratio. The cut-off wavelength can be calculated from the expression $\lambda_{co} = hc / E_{gap}$

The response can be extended to include the visible wavelength range by removing the InP substrate. There has been an intensive effort to develop InGaAs arrays for Low Light Level (LLL) SWIR imaging [40-46]. An example is in astrophysical space based observatories that are very demanding on the detectors due to the very low IR flux levels. Such low flux levels represent the detection of few photons over long integration times and, therefore, require extremely low dark current photodiodes hybridized to a high performance ROIC stage. For such LLL applications there are challenges ahead to further lower noise, reduce pixel size, fabricate larger arrays, achieve higher operating temperatures, and reduce production cost.

The spectral response of InGaAs diodes at room temperature is in the 0.9 – 1.67 μm wavelength range which matches with the ambient night glow spectrum. Imaging under such low light conditions requires that the noise of the detector be extremely low. A significant portion of the noise is contributed by the dark current of the InGaAs detector and the readout noise. Dark current consists of unwanted thermally generated carriers that can cause the detector to produce a random varying output signal.

It is associated with interfacial, diffusional, G-R, and tunneling currents. The temperature dependence of the dark current is primarily due to the intrinsic carrier concentration which depends exponentially on the temperature. The dark current of the detector can be reduced through appropriate fabrication processes and device design. The impact of dark current noise as a function of read noise is shown in Figure 5.1, where the curves for different pixel pitch map the dark current noise into an equivalent read noise.

For a given read noise, the required dark current density increases as the pixel pitch is decreased. The challenge is to maintain a low dark current density as the pixel pitch is reduced. Simultaneously, the challenge for the read out circuit is to reduce the read noise. If the limitation is due to the detector and its noise level overwhelms the source signal, the solution may be to use an external illuminator or cool the detector. The choice of either solution will depend on a tradeoff between size, weight, and power requirements (SWaP).

As mentioned above, the dark current of the detector can be reduced through appropriate fabrication processes and device designs. By focusing on the growth conditions for the InGaAs absorption layer, heterointerfaces and the passivation layer, researchers have been able to demonstrate dark current density below 1.5 nA/cm2 at 77°C for 15 μm pitch arrays as shown Figure 24.

In scaling to small pixel pitch, further effort is continued to develop wafer processing parameters and methods that reduce surface related perimeter effects and enable small pixel pitch InGaAs detectors with dark current densities comparable to large (25 μm) pixel pitches detectors [5-6]. Figure 26 presents a plot of dark current density, measured at 20 °C, for eight different, 300 x 10 pixel test arrays distributed across a 3" wafer. The average dark current density at -100 mV was 2.95 nA/cm2.

Further effort is underway to demonstrate large format (>1Kx1K) and small pixel (<20μm) InGaAs focal plane arrays (FPAs) for a variety of low light level (LLL) imaging applications such as night vision. These applications demand extremely low detector dark current and Si read-out integration circuit (ROIC) noise [47].

Recent work [47] has demonstrated significant progress in InGaAs detector array development on a 4" wafer as shown in figure 27; and also reducing dark current density for 10-20μm pixel arrays, (3) developing sub-10μm pixel array technology and demonstrating the feasibility of making 5μm pixel arrays, and (4) reducing the capacitance of small pixels [47]. Figure 28 demonstrates recent results for spectral quantum efficiency (QE) as a function of wavelength measured on backside illuminated InGaAs photodiodes test array at different temperatures demonstrating Visible-Near IR response with InP substrate removed [47].

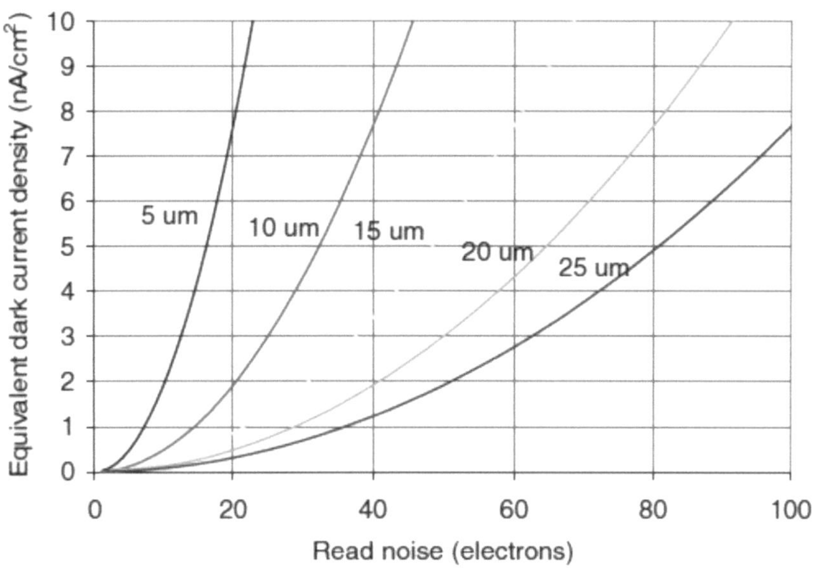

Figure 24. Dark current density versus read noise for different pixel pitches [44].

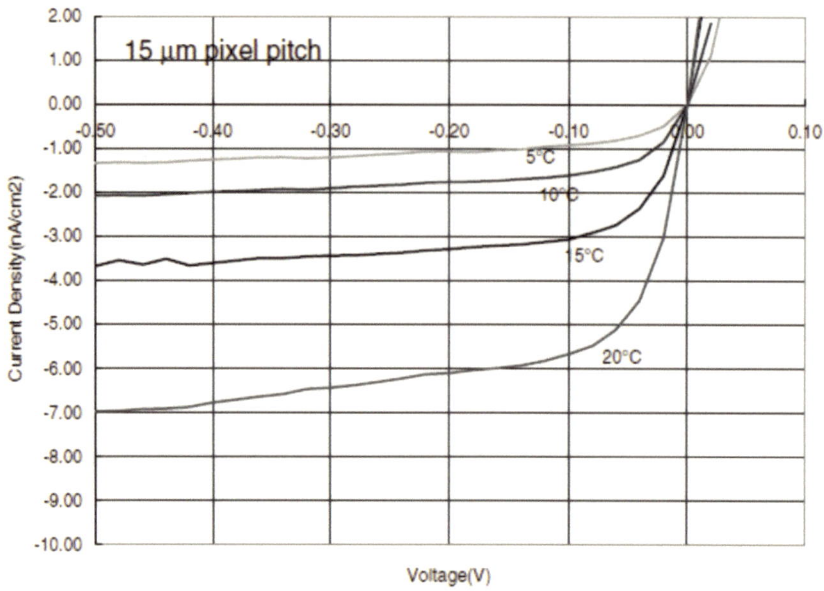

Figure 25. Dark current density at different temperatures using test structures on the wafer. Test arrays have 225 pixels (15 µm pitch) and the guard ring is not biased [45].

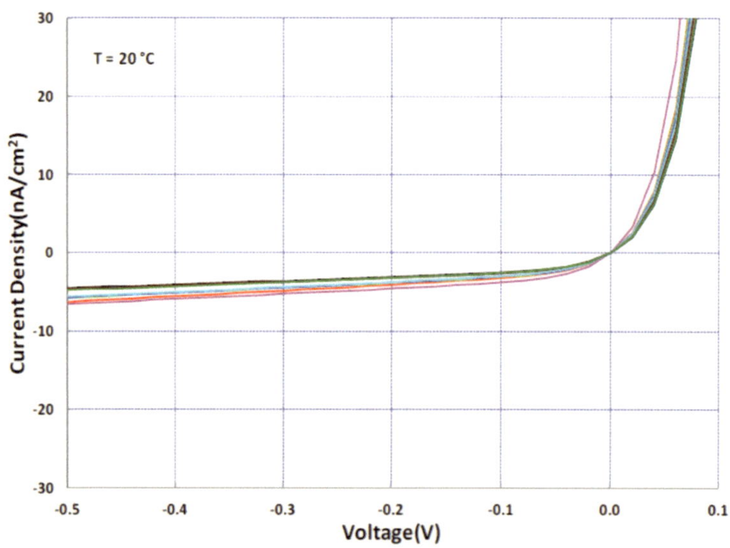

Figure 26. Experimental results for InGaAs test array demonstrating dark current density for eight separate 300 x 10, 15 µm pitch pixel test arrays measured across a wafer. The average dark current density for the test arrays at 100 mV reverse bias is 2.95 nA/cm2 at 20 °C [45].

Figure 27. Experimental 1280x1024/15μm arrays on 4" wafer surrounded by various test mini-arrays with pitch sizes of 5-20μm [47].

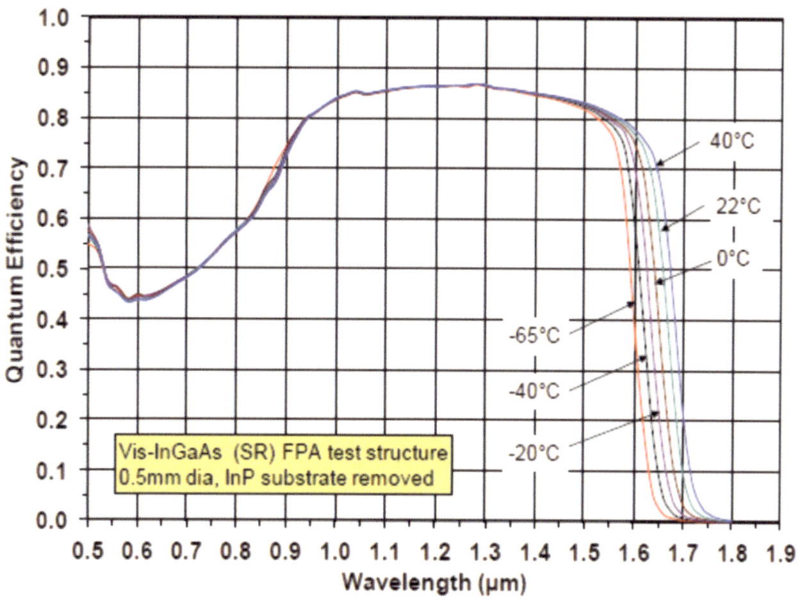

Figure 28. Spectral QE vs. wavelength at different temperatures measured for backside illuminated InGaAs photodiodes test array demonstrating Visible-Near IR response with InP substrate removed [47].

8. Nanostructured detector technology for MWIR and LWIR bands

EO/IR Sensors and imagers using nanostructure based materials are being developed for a variety of Defense Applications. In this section, we will present recent work under way for development of next generation carbon nanostructure based infrared detectors and arrays. We will discuss detector concepts that will provide next generation high performance, high frame rate, and uncooled nano-bolometer for MWIR and LWIR bands [52-55]. The critical technologies being developed include carbon nanostructure growth, characterization, optical and electronic properties that show the feasibility for IR detection. Experimental results on CNT nanostructures will be presented. Further discussion will be provided for the path forward to demonstrate enhanced IR sensitivity and larger arrays.

The microbolometer based on Si-MEMS device structure has been under development for over 20 years with support from DARPA and the US Army. Two most common Si-MEMS based structures utilize VOx and amorphous silicon based technologies. Several companies such as BAE systems and DRS Technologies are developing and producing 17 micron unit cell 640x 480 and larger arrays using VOx [48-51]. Similarly, L3Communications and other groups are developing and producing 640x480 with 17 micron unit cell using amorphous-Silicon technology [50-51].

We will discuss the use of carbon nanostructures for use as the high performance bolometric element of the MWIR and LWIR bands. As part of this effort, we are exploring development of smaller unit cell bolometer.i.e. 5-10 micron unit cell, with higher TCR and higher frequency response in the 1 to 10 KHz range. The feasibility of such an array can open up a larger number of defense and commercial applications. This section will discuss the efforts under way to explore these possibilities.

8.1. Design and modeling of CNT-based bolometer characteristics

To optimize bolometer design, we need to consider several physical key phenomena. From a fundamental point of view, we take the absorbing material to have an extremely large response to infrared radiation. The phonon modes of the material need to be able to easily couple to infrared radiation. Furthermore, once this coupling has been achieved, the absorbed radiation should greatly increase the population of the phonon modes thereby significantly increasing the lattice temperature [52-53].

On a macroscopic scale, this large temperature increase is typically described in terms of a large thermal resistance. The higher the thermal resistance of the material, in general, the higher the resulting temperature will be after absorbing IR. Of course, higher thermal resistance may also give rise to larger thermal noise. This mitigating factor must be balanced with signal response in order to optimize IR sensitivity, or the minimum detectable IR signal.

In summary, for DC operation, we work to maximize the thermal resistance R_{th} while achieving an acceptable noise, which maximizes the minimum signal that we can detect. In addition to DC operation, we try to maximize the thermal frequency response of the bolometer. This

http://dx.doi.org/10.5772/58349

requires that we minimize the thermal capacitance C_{th} of the device and thereby minimize the thermal $R_{th}C_{th}$ time constant for the absorbing material.

Temperature Dependent Electrical Characteristics: In addition to considering the DC and transient thermal characteristics of the absorbing material, we need to optimize the electrical response as well. To achieve this, we associate the electrical response with the thermal response by considering the temperature dependent voltage-current characteristics of the electrical material.

A good response is obtained by utilizing a material that has an effectively large variation of electrical resistance with temperature. However, at the same time we want the material to have a temperature coefficient of resistance that is relatively independent of the absolute value of the resistance itself. Therefore, we look at materials that have an exponential relationship between electrical resistance and temperature. For such materials, the Temperature Coefficient of Resistance (TCR) is not strongly dependent on the absolute value of the electrical resistance of the material itself.

8.1.1. Calculating the thermal response of the CNT bolometer film

In this work we are assessing the possibility of designing a bolometer using carbon nanotubes as both the IR absorbing material and the electrical response material. Thus, our aim is to first determine the thermal response of the bolometer absorber that is composed of CNTs, and then determine how the electrical characteristics of the CNT material depend on its changes in temperature after IR absorption.

While there are numerous geometries of CNT based material we can consider, for this work we will focus on an absorbing material composed of a CNT film. The film is taken to consist of a random placement of CNTs that is two nanotubes thick.

Heat Flow Equation: To determine the temperature of the material in the presence of IR radiation, we start with the heat flow equation. This is a partial differential equation relating the time rate of change in temperature to the position and the rate of net heat that is absorbed by the material as a function of time and position.

$$Cv\frac{\partial T}{\partial t} = \kappa \nabla^2 T + H$$

In the equation C_v is the thermal capacitance (joules/degree-unit cm^3) and κ is the thermal diffusion coefficient in watts/degree-cm). H is the net power absorbed by the material in watts per unit volume. To solve this equation for the CNT bolometer, we have to first determine C_v and κ for the CNT and the CNT film.

Thermal Capacitance of CNT Absorber: To determine the heat capacity of a carbon nanotube, we first determine the internal vibrational energy of the CNT, and then take the derivative with respect to temperature. The internal energy is found by determining the energy of each vibrational mode, multiplying by the probability that the mode is populated using Bose-Einstein statistics, and then summing over all of the allowed modes.

The number of allowed modes will depend on the diameter and wrapping angle of the CNTs present, so we take a statistical sample. Multiplying the individual CNT heat capacity by the number of CNTs in the film provides a reasonable value for the heat capacity of the film. After following this procedure and inserting numerical values for physical constants, we arrive at the following average numerical value for thermal capacity of a CNT C_{vt}, where the length L is in microns and the diameter d is in nanometers

$$C_{vt} \sim 1.4 \times 10^{-18} (L)(d)$$

CNT Thermal Diffusion Coefficient: In addition to the thermal capacitance, we need to determine the thermal diffusion coefficient, and eventually the thermal resistance of a single CNT. Experiments on isolated CNTs have fit the coefficient of thermal diffusion to data obtaining the following expression [52-53]:

$$\kappa(L, T) = \left\{ 3.7x10^{-7}T + 9.7x10^{-10}T^2 + \frac{9.3}{T^2}\left[1 + \frac{0.5}{L}\right] \right\}^{-1}$$

From κ we can obtain the thermal resistance R_T of a single CNT using the following definition:

$$R_T = \frac{4L}{\kappa \pi d^2}$$

Where L, d are the length and diameter of the CNT

Using average values for CNTs gives the following numerical value for the thermal resistance of a CNT

$$R_T \sim 5 \times 10^8 (L/d^2)$$

Where R_T is in units of degrees K/Watt, L is in units of microns and d is in units of nanometers. So a tube that is one micron long and one nanometer in diameter will have a thermal resistance R_T of approximately 5×10^8 K/W.

Net IR Radiation Power Absorbed: Now that we have the thermal diffusion and capacitance we are almost ready to begin solving the above heat flow equation to determine the temperature of the bolometer. However, before doing so, we need to determine H, the net IR power absorbed by the bolometer. We determine this power using the Stefan-Boltzmann Law of blackbody radiation, which relates the net power absorbed to the temperatures of the subject and the bolometer using the following expression:

$$H_{net} = \sigma A \varepsilon (T_{obj}^4 - T_b^4)$$

Where H_{net}, σ, A, ε, T_{obj} and T_b are heat absorbed by the bolometer, the Stefan-Boltzmann constant, cross-sectional area, emissivity, object of interest temperature and bolometer absorber temperature, respectively. Figure 1 below shows the net IR power absorbed by the absorber as a function of bolometer temperature for radiating objects at 20°C and 36.5°C. Cooling the bolometer by 30°C below room temperature allows for significantly more power to be absorbed, which can give rise to a much stronger signal.

Calculating the Bolometer Temperature Distribution: Using the aforementioned expressions for C_{vt}, K and H, we expand on previous work and convert the heat flow equation above into a

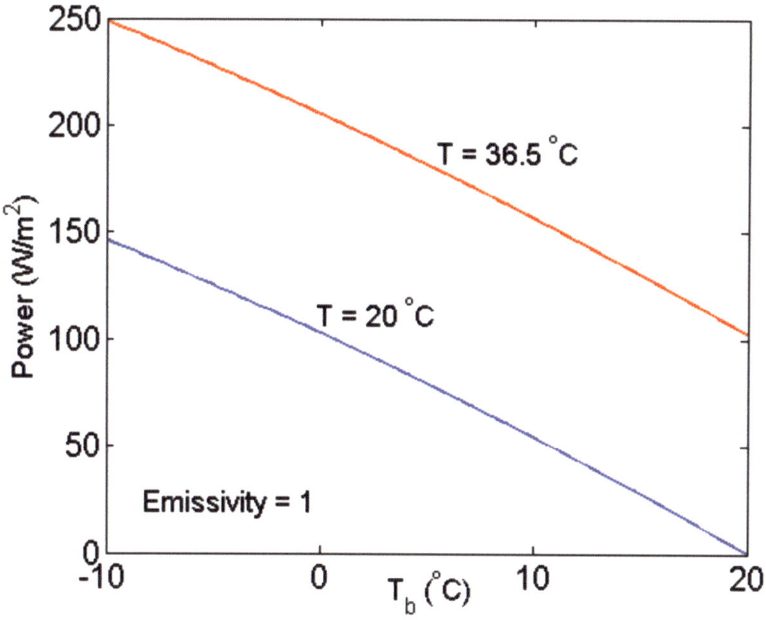

Figure 29. Net power received by bolometer as a function of bolometer temperature.[54]

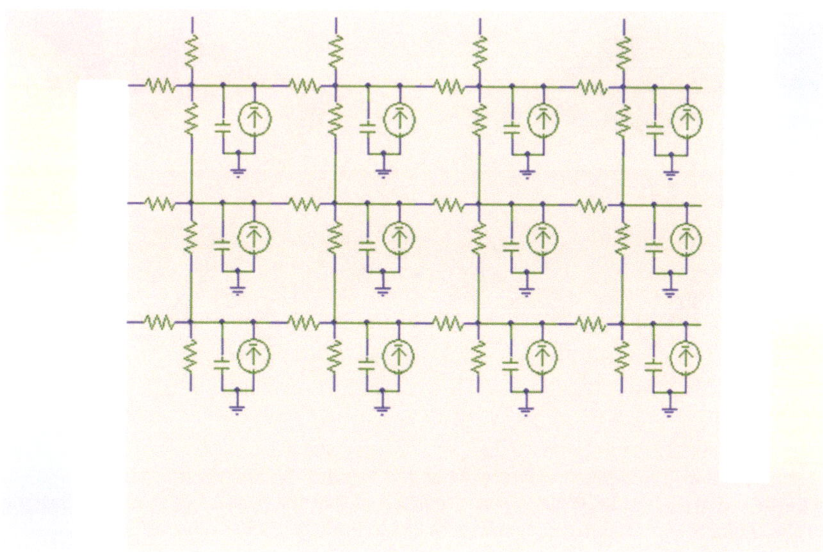

Figure 30. Illustration of thermal network superimposed on bolometer for calculating temperature map of CNT bolometer absorber.[55]

thermal network, illustrated in Figure 2 [6]. In actuality, there are thousands of nodes in the network for which we calculate the temperature for each.

In Figure 30, each resistor represents the thermal resistance of a CNT in series with the thermal resistance between adjacent CNTs. In addition, the capacitors represent the thermal capacity of a CNT, while the current sources represent the net IR radiation absorbed by each CNT. This thermal network contains thousands of nodes, and there is an equation relating the thermal resistance, capacitance and net power for each node. This system of equations is then solved for the temperature as a function of position and time throughout the bolometer absorber [54].

Results of these calculations for are shown in Figure 31 for different types of CNT networks. Here, we assumed that the net absorbed power is 1 nW, and the pixel is tightly packed with the CNTs. The entire pixel's temperature map is obtained with a 100×100 temperature resolution. For the tubes, we used two different thermal resistance values: 5×10^8 and 1×10^9 K/W.

As expected, higher the thermal resistance, higher the temperature difference from the ambient. We note that in general thermal resistance also rises with increasing temperature, resulting in further heating of hot spots compared to the case that this dependency is ignored. The temperature gradient of the contacts legs connecting the film to the readout IC (ROIC) is clearly shown.

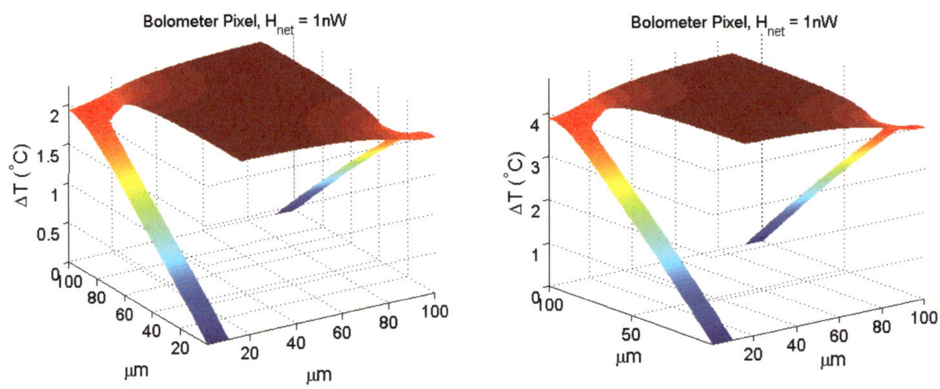

Figure 31. Temperature map of a bolometer pixel when the net absorbed power *H* is 1 nW, and the CNT thermal resistance is 5×10^8 K/W (left) and 1×10^9 K/W (left). We assumed that the pixel is tightly packed with the CNTs. The temperature gradient of the contacts legs connecting the film to the ROIC is clearly shown [54]

8.1.2. Calculating the electrical response of the CNT film

To read the temperature that the bolometer pixel reaches after an exposure to infrared radiation, one needs to measure the electrical resistance of the pixel. By comparing this resistance to a look-up table or using the a-priori knowledge of temperature coefficient of

http://dx.doi.org/10.5772/58349

resistance, the pixel temperature can be determined. Therefore, in addition to having a large thermal resistance which translates into higher temperature rises, a large temperature coefficient of electrical resistance (TCR) is desirable to achieve a higher temperature resolution.

Here TCR is defined as the change in electrical resistance per degree Kelvin divided by the absolute electrical resistance measured at the quiescent point, as follows:

$$TCR = \frac{1}{R_e}\frac{dR_e}{dT}$$

Thus, the pixel electrical resistance after it reaches a temperature that is ΔT above its ambient becomes $R_e(T) = R_e(To)(1+TCR)$. Using this relationship, the pixel temperature is calculated.

To obtain a high temperature resolution, a large change in electrical resistance is needed upon heating. To achieve this, a substantial increase either in electron concentration or velocity (for a given electric field) is necessary. And to this end, materials with junctions where thermionic emission or tunneling are the electrical current bottlenecks offer a good solution. As the tunneling current exponentially rises with temperature, the effective change in their electrical resistance due to temperature becomes large compared to those observed in bulk materials where the change is proportional T^{γ} and γ is generally < 2.

Here, a film of CNTs is proposed as the bolometer pixel material, since it is likely to have large thermal resistance and TCR values simultaneously. Both of these favorable properties are partially owed to the junctions between the tubes. As the electrical current flows along the mat, it needs to jump from one tube to the next where they intersect.

At this intersection, the carriers see a potential barrier that they need to tunnel through which gives rise to exponential increase in current upon heating. Assuming that the electron transport across this barrier is governed by a Fowler-Nordheim-type tunneling or thermionic emission, the expected TCR values can be calculated using the following expression:

$$I = q\int_0^{\infty} T_t(E)v(E)DOS(E)f(E, T)dE$$

Where T_t, v, DOS and f, are the transmission coefficient, thermal velocity, density of states and distribution function, respectively for electrons in the CNT. We perform this calculation a function of barrier height and electric field. The results are shown in figure 4. The figure on the left shows a contour plot of the theoretical values of the TCR for a CNT film.

The TCR is plotted as a function of electric field between the tubes and the barrier height. Theoretical calculations predict an extremely large TCR, which can be attributed to the relatively large barrier height between adjacent CNTs. If a lower barrier height is assumed, on the order of 0.06eV, then a TCR of approximately 2.5% is obtained. It is also worth pointing out that such a large TCR comes at the price of extremely low output currents. The bolometer current densities, as a function of barrier height and electric field are shown in Figure 32, on the right.

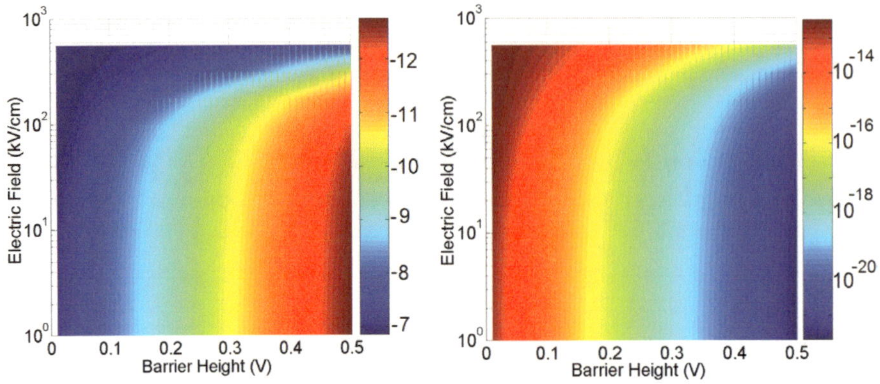

Figure 32. Left figure is a contour plot of TCR versus electric field and barrier height between CNTs of the film. The right figure shows the bolometer current also as a function of electric field and barrier height. The scales are the color bars on the right of each contour plot in units of %TCR and amperes, respectively.

8.1.3. CNT growth and charcterization

In this section, we will discuss growth and characterization of carbon nanotubes with single wall (SWCNT) and multiwall (MWCNT) for use as the high performance bolometric element for development of MWIR and LWIR sensitive detector elements.

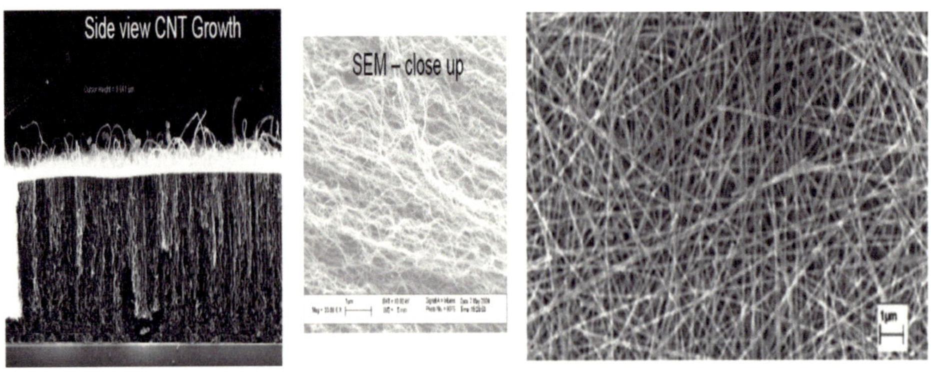

Figure 33. Growth of multiwall CNT forest with the ability to separate the form growth substrate with good length / diameter uniformity and the MWCNT released from the template.[55]

Figure 33 presents growth of dense oriented multi-walled CNT "forest like growth". The figure shows the CNT growth can be easily separated from the growth substrate. We have shown good length/diameter uniformity. Further work on the growth optimization is underway.

Figure 34 shows the prototype fixture to evaluate the CNT films for bolometric application. This fixture is being used for quick evaluation of both electrical and optical characteristics of the CNT samples. The figure also shows the preliminary results of reflectivity measurements for SWCNT and MWCNT samples with various sample treatments. We have carried out some preliminary measurements of TCR on CNT samples.

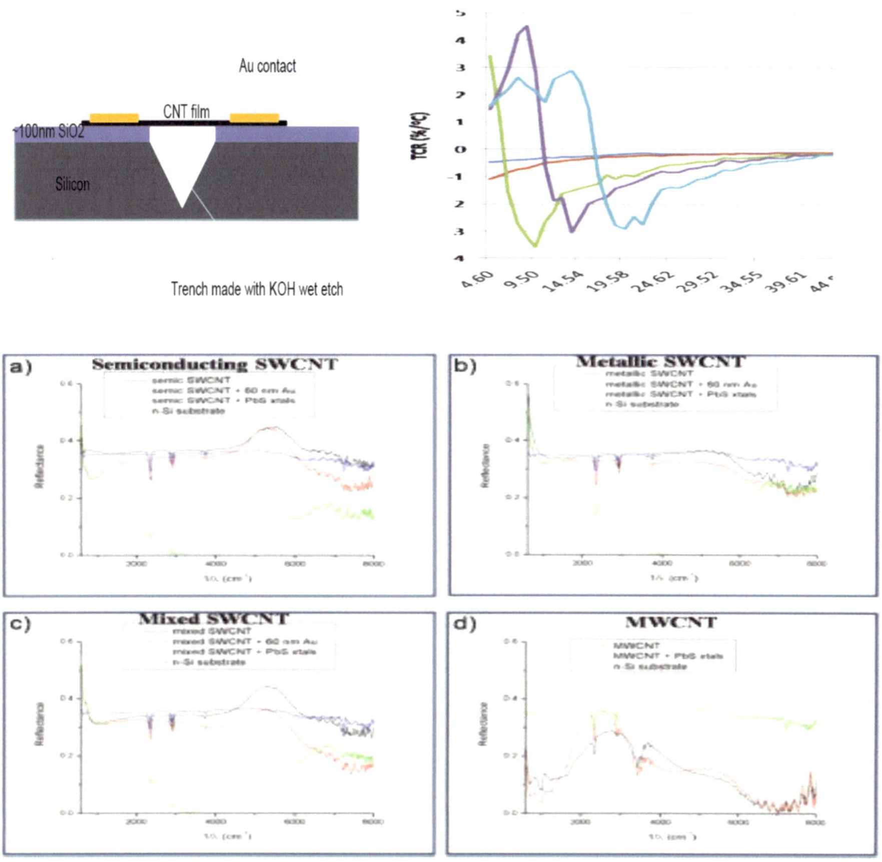

Figure 34. The composite figure shows prototype CNT film bolometer test fixture to evaluate the CNT film quality. Some preliminary data on reflectivity measurements for SWCNT and MWCNT are shown along with preliminary results on TCR measurements.[55]

Figure 35 shows scanning electron microscopy SEM. images of representative MWCNT films in the unsuspended left. and suspended.right. forms, respectively. Unlike their SWCNT counterparts, the MWCNT films contain substantial uncovered substrate areas. In addition, some minor deformation of recess is visible on suspended MWCNT films, which is similar to the SWCNT film case in the same thickness range. Figure 35.b includes a transmission electron microscopy.TEM. image of a representative individual MWCNT, which has a large hollow center of approximately 10–11 nm in diameter and contains approximately 40–50 CNT shells.

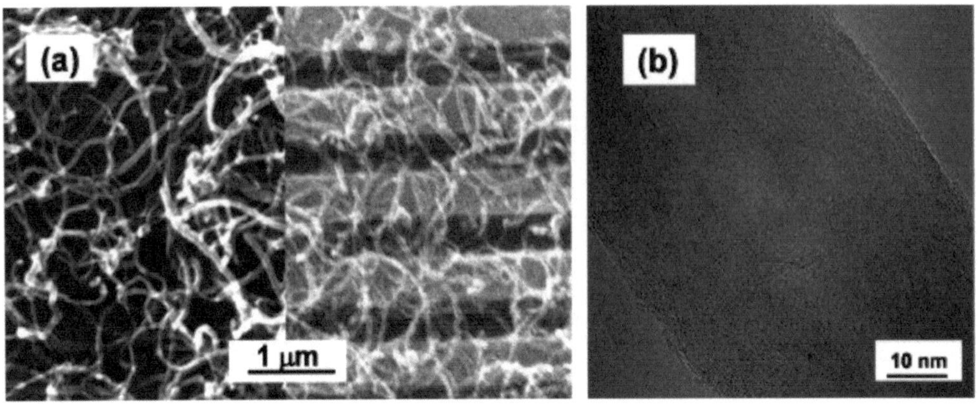

Figure 35. SEM images of unsuspended.left (a). and suspended.right (b). MWCNT films. A TEM image of a representative MWCNT. The shell number is estimated to be.40–50 for the MWCNTs [56]

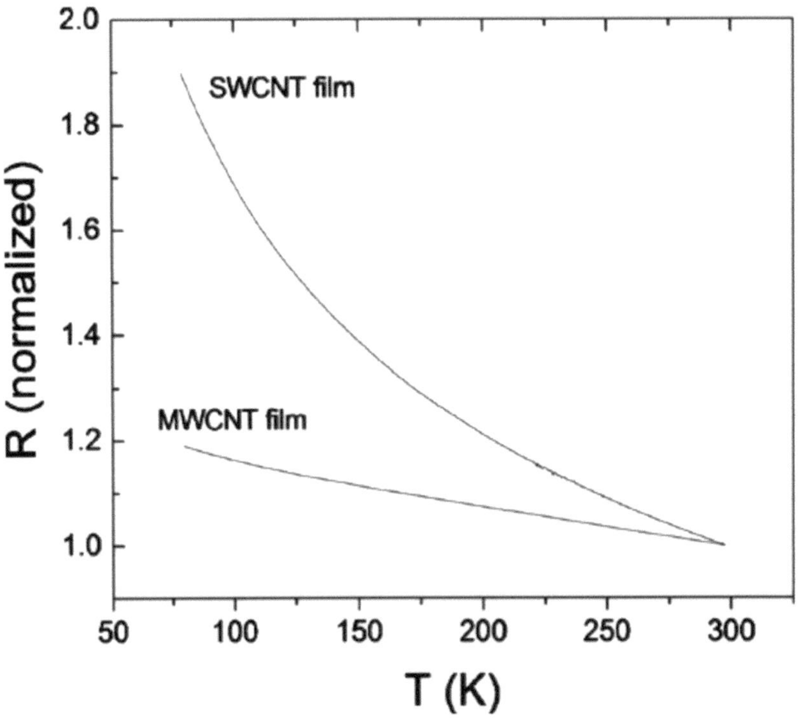

Figure 36. Resistance versus temperatures curves of SWCNT films and MWCNT films.[56]

All MWCNT films studied in this work [56] show semiconductive resistance-temperature.R-T. behaviors and a representative curves is depicted in Figure 36.. Nevertheless, the increase in the resistivity of MWCNT films is much less than that of SWCNT films with decreasing temperature, as shown in Figure 36. This is not unexpected considering a much smaller band gap in MWCNTs. The reduced temperature dependence also implies smaller TCR absolute value in MWCNTs. For example, the TCR absolute value at room temperature for MWCNT films is about 0.07%/K in contrast to.0.17%/K for SWCNT films. R-T curve after suspending the MWCNT film has been also measured.

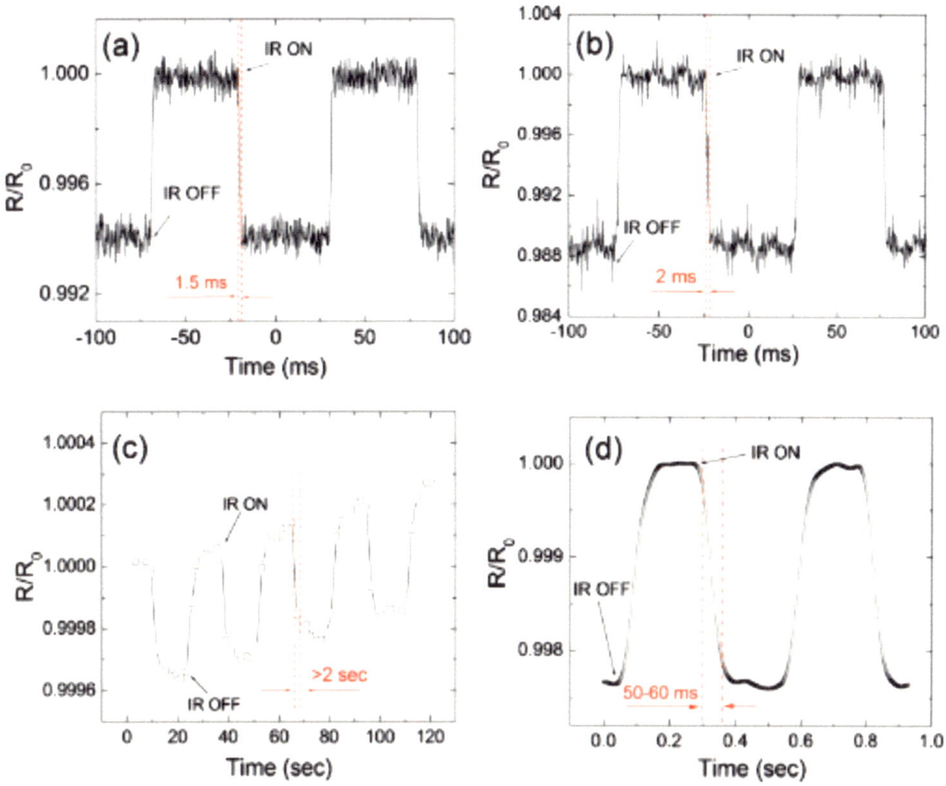

Figure 37. Photoresponse of unsuspended and suspended CNT films.(a) Unsuspended MWCNT film, f=10 Hz, in IR.3 mW/mm2 ;.(b). suspended MWCNT film, f=10 Hz, in IR.3 mW/mm2 ; (c) unsuspended SWCNT film, f=1/30 Hz, in IR.3.5 mW/mm2 ; and (d) suspended SWCNT film, f =2 Hz, in IR.3.5 mW/mm2 [56].

Figure 37 compares the Photoresponse R/R0 of MWCNT films in unsuspended (a) and suspended (b) cases, where R_0 is the sample resistance before IR radiation was turned on and the change in the resistance caused by IR radiation is defined as.R=R-R_0. For comparison, the results of their SWCNT counterparts are also included in Figure 37 (c).unsuspended and Fig. 37(d). suspended..

Two major differences are visible between MWCNT and SWCNT films, a significantly higher.R/R_0 and a much shorter response time in the cases of MWCNT. The.R/R_0 for MWCNT samples is typically in the range of a few percent, which is more than one order of magnitude higher than that of suspended SWCNT films and two orders of magnitude higher than the unsuspended SWCNT films at a comparable IR power. Considering a lower TCR absolute value in MWCNTs, the much enhanced Photoresponse of MWCNT films should be attributed to the naturally suspended inner CNT shells, which may provide an ideal configuration to enhance the bolometric effect by improving light absorption and reducing thermal link. Physical suspension of the films in both MWCNT.Fig. 37.(b) and SWCNT.Fig. 37(d.) cases results in a further improvement of.R/R_0 as compared to their unsuspended counterparts. The improvement is, however, much more pronounced in suspended cases [56].

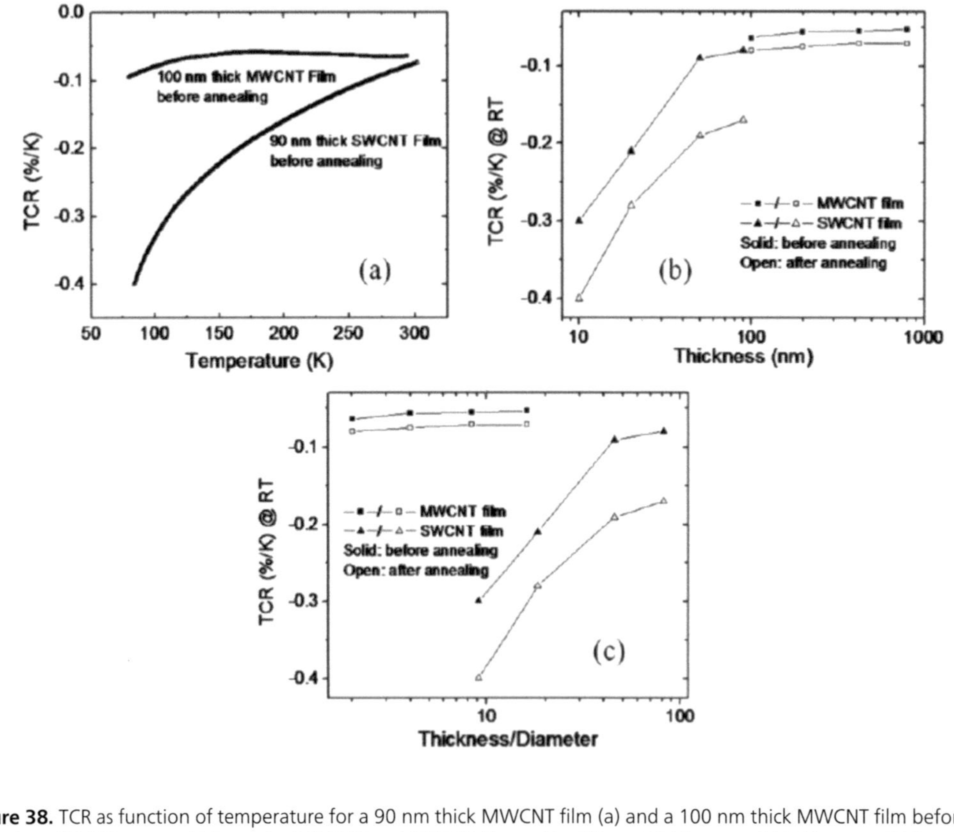

Figure 38. TCR as function of temperature for a 90 nm thick MWCNT film (a) and a 100 nm thick MWCNT film before annealing. (b) TCR versus thickness for MWCNT and SWCNT films with different thicknesses. (c) TCR versus thickness/diameter ratios for SWCNT and MWCNT films [57].

We have also shown the results of TCR as function of temperature in figure 38, for a 90 nm thick MWCNT film (a) and a 100 nm thick MWCNT film before annealing. (b) TCR versus thickness for MWCNT and SWCNT films with different thicknesses. (c) TCR versus thickness/diameter ratios for SWCNT and MWCNT films [57].

We have discussed recent efforts for modeling CNT based bolometer and the experimental work for development of next generation carbon nanostructure based infrared detectors and arrays. Our goal is to develop high performance, high frame rate, and uncooled nano-bolometer for MWIR and LWIR bands. We also discussed CNT growth system and its capability to grow samples of various orientations. We have also presented recent results on SWCNT and MWCNT samples that show promise for use of CNT for developing next generation high performance small pixel bolometer arrays.

9. Summary

In this chapter, we have discussed recent advances in nanostructured based detector technology, materials and devices for optical sensing applications. The chapter has presented an overview of recent work underway on a variety of semiconductors and advanced materials such as GaN, ZnO, Si/SiGe, InGaAs and CNT for optical sensing applications.

Optical sensing technology is critical for defense and commercial applications including optical communication. Advances in optoelectronics materials in the UV, Visible and Infrared, using nanostructures, and use of novel materials such as CNT have opened doors for new approaches to apply device design methodology that are expected to offer enhanced performance and low cost optical sensors in a wide range of applications.

We have covered the UV band (200-400 nm) and address some of the recent advances in nanostructures growth and characterization using GaN/AlGaN, ZnO/MgZnO based technologies and their applications. We have also discussed nanostructure based Si/SiGe technologies (400-1700 nm) that covers various bands of interest in visible-near infrared for detection and optical communication applications. The chapter has also discussed some of the theoretical and experimental results in these detector technologies.

Recent advancements in design and development of CNT based detection technologies have shown promise for optical sensor applications. We have presented theoretical and experimental results on these device and their potential applications in various bands of interest. It is anticipated that the current research and development presented in this chapter will enable a host of new integrated technologies for a variety of defense and commercial applications.

Although numerous research activities are ongoing in the area of Nanoscience and technology, we briefly made comments on such technologies to make readers aware of various research activities.

Acknowledgements

The authors gratefully acknowledge the contributions of the many distinguished scientists in the United States for development of nanotechnology based EO/IR detector technology for optical sensor applications.

Author details

Ashok K. Sood[1], Nibir K. Dhar[2], Dennis L. Polla[3], Madan Dubey[4] and
Priyalal Wijewarnasuriya[4]

1 Magnolia Optical Technologies Inc., Woburn, MA, USA

2 Defense Advanced Project Agency, Arlington, VA, USA

3 College of Science and Engineering, University of Minnesota, Minneapolis, MN, USA

4 Army Research Laboratory, Adelphi, MD, USA

References

[1] Raytheon Vision Systems Spectral Charts, Goleta, CA

[2] N. K. Dhar. " IR Material Research at the Army Research Laboratory" Invited Keynote Paper, Proceedings of SPIE, Volume 6542, 65420C (2007)

[3] J.P. Long, "UV detectors and focal plane array imagers based on AlGaN p-i-n photodiodes", Opto-Electronics Review 10(4), 251-260 (2002)

[4] M. Crawford, "Advances in AlGaN-based deep UV LEDs ", MRS Proceedings, Vol. 831, fall 2004.

[5] M.B. Reine, "Solar-blind AlGaN 256x256 p-i-n detectors and focal plane arrays", Proc. of SPIE, Vol. 6119 (2006)

[6] Ashok K. Sood, Robert A. Richwine, Yash R. Puri, Nibir K. Dhar, Dennis L. Polla, and Priyalal S. Wijewarnasuriya, " Multispectral EO/IR sensor model for evaluating UV, visible, SWIR, MWIR and LWIR system performance" Proceedings of SPIE 7300, 73000H (2009)

[7] Abdiel Rivera, John Zeller, Tariq Manzur, Ashok Sood and Mehdi Anwar, " MOCVD Growth and Characterization of ZnO Nanowire Arrays for UV Detectors" Proceedings of SPIE, Volume 8540, October 2012.

[8] Liang, S., Sheng, S., Liu, Y., Huo, Z., Lu, Y., and Shen, H., "ZnO Schottky ultraviolet photodetectors," J. Cryst. Growth 225, 110-113 (2001).

[9] Zhang, J., Que, W., Jia, Q., Ye, X., and Ding, Y., "Controllable hydrothermal synthesis of ZnO nanowires arrays on Al-doped ZnO seed layer and patterning of ZnO nanowires arrays via surface modification of substrate," Appl. Surf. Sci. 257(23), 10134-10140 (2011).

[10] Lee, C. H., Yi, G. C.., Zuev, Y. M., and Kim, P., "Thermoelectric power measurements of wide band gap semiconducting nanowires,"Appl. Phys. Lett. 94, 22106 (2009).

[11] Falyouni, F., Benmamas, L., Thiandoume, C., Barjon, J., Lusson, A., Galter, P., and Sallet, V., "Metal organic chemical vapor deposition growth and luminescence of ZnO micro- and nanowires," Journal Vac. Sci. Technol. B 87, 1662 (2009).

[12] Jeong, M. C., Oh, B.Y., Lee, W., and Myoung, J. M., "Comparative study on the growth characteristics of ZnO nanowires and thin films by metal-organic chemical vapor deposition (MOCVD)," Journal of. Crystal Growth 268, 149-154 (2004).

[13] Kim, S. W., Fujita, S., and Fujita, S., "ZnO nanowires with high aspect ratios grown by metal-organic chemical vapor deposition using gold nanoparticles," Appl. Phys. Lett. 86,153119 (2005).

[14] Lee, W., Jeong, M. C., and Myoung, J. M., "Catalyst-free growth of ZnO nanowires by metal-organic chemical vapor deposition (MOCVD) and thermal evaporation," Acta Mat. 52, 3949-3957 (2004).

[15] Liou, S. C., Hsiao, C. S., and Chen, S. Y., "Growth behavior and microstructure evolution of ZnO nanorods grown on Si in aqueous solution," Journal of. Crystal. Growth 274, 438 (2005).

[16] Dong, J. W., Osinski, A., Hertog, B., Dabiran, A. M., Chow, P. P., Heo, Y. W., Norton, D. P, and Pearton, S. J., "Development of MgZnO-ZnO-AlGaN heterostructures for ultraviolet light emitting applications," J. Electron. Mat. 34, 416-423 (2005).

[17] Rivera, A., Zeller, J., Sood, A.K., and Anwar, A. F. M., "A Comparison of ZnO Nanowires and Nanorods Grown Using MOCVD and Hydrothermal Processes," J. Electron. Mat. 42, 894-900 (2013).

[18] Ha, B., Ham, H., and Lee, C. J., "Photoluminescence of ZnO nanowires dependent on O_2 and Ar annealing," Phys. Chem. Solids 69, 2453-2456 (2008).

[19] Djurišić, A. B., Ng, A.M.C., and Chen, X.Y., "ZnO nanostructures for optoelectronics: Material properties and device applications," Progress Quantum Electronics 34, 191-259 (2010)

[20] Mehdi Anwar, Abdiel Rivera, Anaz Mazady, Hung Chou, John Zeller and Ashok K. Sood, "ZnO Solar Blind Detectors: from Material to System", Proceedings of SPIE Volume 8868, 8868B (2013).

[21] Zhong Lin Wang, Guang Zhu, Ya Yang, Sihong Wang and Caofeng Pan, " Progress in Nanogenerator for Portable Electronics" Materials Today Volume 15, Number 12 December 2012

[22] Wenzhuo Wu, Xiaonan Wen and Zhong Lin Wang, " Texel- Addressable Matrix of Vertical-Nanowire and Adaptive Tactile Imaging" Science, Volume 340, 24 May 2013.

[23] R.D. Dupuis, H.J. Ryou and D. Yoder. "High-performance GaN and AlGaN ultraviolet avalanche photodiodes grown by MOCVD on bulk III-N substrates", Proc. of SPIE, Vol. 6739 (2006)

[24] S.C. Shen, Y. Chang, J. B. Limb, J.H. Ryou, P. D. Yoder and R.D. Dupuis, " Performance of Deep UV GaN Avalanche Photodiodes Grown by MOCVD", IEEE Photonics Technology Letters, Volume 19, Number 21, November 2007.

[25] Y. Zhang, S.C. Shen, H. J. Kim, S. Choi, J. H. Ryou, R.D. Dupuis and B. Narayan, " Low-Noise GaN Ultraviolet p-i-n photodiodes on GaN Substrates" Applied Physics Letters, 94, 221109 (2009).

[26] Ashok K. Sood, Robert A. Richwine, Roger E. Welser, Yash R. Puri, Russell D. Dupuis, Mi-Hee Ji, Jemoh Kim, Theeradetch Detchprohm, Nibir K. Dhar and Roy L. Peters, " Development of III-N UVAPDs for Ultraviolet Sensor Applications" Proceedings of SPIE Volume 8868, 88680T (2013)

[27] Ashok K. Sood, Robert A. Richwine, Yash R. Puri, Russell D. Dupuis, Nibir K. Dhar and Raymond S. Balcerak "Development of GaN/AlGaN APD's for UV Imaging Applications" Proceedings of SPIE 7780, 77800E (2010)

[28] J.P. Long, "UV detectors and focal plane array imagers based on AlGaN p-i-n photodiodes", Opto-Electronics Review 10(4), 251-260 (2002)

[29] M. Crawford, "Advances in AlGaN-based deep UV LEDs ", MRS Proceedings, Vol. 831, fall 2004.

[30] M.B. Reine, "Solar-blind AlGaN 256x256 p-i-n detectors and focal plane arrays", Proc. of SPIE, Vol. 6119 (2006)

[31] Ashok K. Sood, Robert A. Richwine, Yash R. Puri, Nibir K. Dhar, Dennis L. Polla, and Priyalal S. Wijewarnasuriya, " Multispectral EO/IR sensor model for evaluating UV, visible, SWIR, MWIR and LWIR system performance" Proceedings of SPIE 7300, 73000H (2009)

[32] S.C. Shen, Y. Chang, J. B. Limb, J.H. Ryou, P. D. Yoder and R.D. Dupuis, " Performance of Deep UV GaN Avalanche Photodiodes Grown by MOCVD", IEEE Photonics Technology Letters, Volume 19, Number 21, November 2007.

[33] Y. Zhang, S.C. Shen, H. J. Kim, S. Choi, J. H. Ryou, R.D. Dupuis and B. Narayan, " Low-Noise GaN Ultraviolet p-i-n photodiodes on GaN Substrates" Applied Physics Letters, 94, 221109 (2009).

[34] B.-Y. Tsaur, C. K. Chen and S. A. Marino, "Long-wavelength Ge_x Si_{1-x} /Si heterojunction infrared detectors and focal plane array," *Proc. SPIE 1540*, 580-595, 1991.

[35] H. Wada, M. Nagashima, K. Hayashi, J. Nakanishi, M. Kimata, N. Kumada and S. Ito, "512 x512 element GeSi/Si heterojunction infrared focal plane array," *Proc. SPIE* 3698, 584-595, 1999.

[36] A. K. Sood, R. A. Richwine, Y. R. Puri, N. DiLello, J. L. Hoyt, N. Dhar, R. S. Balcerak, and T. G. Bramhall, "Development of SiGe Arrays for Visible-Near IR Imaging Applications," *Proc. SPIE* 7780,77800F, 2010.

[37] V.T. Bublik, S.S. Gorelik, A.A. Zaitsev and A.Y. Polyakov, "Calculation on the Binding Energy of Ge-Si Solid Solution," *Phys. Status Solidi* 65, K79-84, 1974.

[38] A. K. Sood, R. A. Richwine, Y. R. Puri, N. DiLello, J. L. Hoyt, T. Akinwande, N. K. Dhar, R. S. Balcerak, and T. G. Bramhall, "Characterization of SiGe detector arrays for Visible-Near IR Imaging Sensor Applications," *Proc. SPIE* 8012,801240, 2011.

[39] Ashok K. Sood, Robert A. Richwine, Gopal Pethuraja, Yash R. Puri, Je-Ung Lee, Pradeep Haldar and Nibir K. Dhar, " Design and Development of Wafer-Level Short Wave Infrared Micro-Camera" Proceedings of The SPIE Volume 8704, 870439 (2013).

[40] R.J. van der A, R.W.M. Hoogeveen, H.J. Spruijt, and A.P.H. Goede, "Low noise InGaAs infrared (1.O-2.4tm) Focal Plane Arrays," *Proc. SPIE* 2957, 54-65, 1997.

[41] D. Acton, M. Jack, and T. Sessler, "Large format short-wave infrared (SWIR) focal plane array (FPA) with extremely low noise and high dynamic range," *Proc. SPIE* 7298, 72983E, 2009.

[42] B. M. Onat, W. Huang, N. Masaun, M. Lange, M. H. Ettenberg, and C. Dries, "Ultra low dark current InGaAs technology for focal plane arrays for low-light level visible-shortwave infrared imaging," *Proc. SPIE* 6542, 65420L, 2007.

[43] J. Boisvert, T. Isshiki, R. Sudharsanan, P. Yuan, and P. McDonald, "Performance of very low dark current SWIR PIN arrays," *Proc. SPIE* 6940, 69400L, 2008.

[44] M. MacDougal, J. Geske, J. Wang, and D. Follman, "Short-wavelength infrared imaging using low dark current InGaAs detector arrays and vertical-cavity surface-emitting laser illuminators," Optical Engineering 50(6), 061011, 2011.

[45] A. D. Hood. "Large InGaAs Focal Plane Arrays for SWIR Imaging" Proceedings of SPIE, Volume 8353, 83530A, 2012.

[46] H. Yuan, M. Meixell, J. Zhang, P. Bey, J. Kimchi, and L.C. Kilmer "Low Dark Current Small Pixel Large Format InGaAs 2-D Photodetector array Development" Proceedings of SPIE, Volume 8353, 835309, 2012.

[47] H. Yuan, G. Apgar, J. Kim, J. Laquindanum, V. Nalavade, P. Beer, J. Kimchi, and T. Wong, "FPA development: from InGaAs, InSb, to HgCdTe," *Proc. SPIE* 6940, 69403C, 2008.

[48] R. Blackwell, D. Lacroix, T. Bach, et. al " 17 micron microbolometer FPA Technology at BAE Systems" Proceedings of SPIE, Volume 7298,72980P (2009).

[49] C. Li, G. Skidmore, C. Howard, E. Clarke and J. Han, "Advancement in 17-micron pixel pitch uncooled focal plane arrays" Proceedings of SPIE, Volume 7298, 72980S (2009).

[50] T. Schimert, C. Hanson, J. Brady, et. al. " Advanced in small-pixel, large-format alpha-silicon bolometer arrays", Proceedings of SPIE, Volume 7298, 72980T (2009)

[51] C. Trouilleau, B. Fieque, S. Noblet, F. Giner et.al. "High Performance uncooled-amorphous silicon TEC less XGA IRFPA with 17 micron pixel pitch" Proceedings of SPIE Volume 7298, 72980Q, (2009).

[52] E. Pop, D. Mann, Q. Wang, K. Goodson, H. Dai, Thermal Conductance of an Individual Single-Wall Carbon Nanotube above Room Temperature Nano Letters 6, 96 (2006)

[53] A. Akturk, N. Goldsman, G. Metze, "Self-consistent modeling of heating and MOSFET performance in 3-D integrated circuits," IEEE Trans. on Elect. Dev. 52 (11): 2395-2403 (2005).

[54] Ashok K. Sood, E. James Egerton, Yash R. Puri, Gustavo Fernandes, Jimmy Xu, Akin Akturk, Neil Goldsman, Nibir K. Dhar, Madan Dubey, Priyalal S. Wijewarnasuriya and Bobby I Lineberry, " Design and Development of CNY based Micro-bolometer for IR Imaging Applications" Proceedings of SPIE, Volume 8353, 83533A, May 2012

[55] Gustavo Fernandes, Jin Ho Kim, Jimmy Xu, Ashok K. Sood, Nibir K. Dhar and Madan Dubey, " Unleashing Giant TCR from Phase –Changes in Carbon Nanotube Composites" Proceedings of SPIE Volume 8868, 88680S, September 2013

[56] Rongtao Lu, Jack J. Shi, F. Javier Baca and Judy Z. Wu, " High Performance Multiwall Carbon Nanotube Bolometer" Journal of Applied Physics, 108, 084305 (2010).

[57] Rongtao Lu, Rayyan Kamal and Judy Z. Wu, " A comparative study of 1/f noise and temperature coefficient of resistance in multiwall and single-wall carbon nanotube bolometers" nanotechnology, Volume 22, 265503 (2011).

Chapter 10

Dipping Deposition Study of Anodized-Aluminum Pressure-Sensitive Paint for Unsteady Aerodynamic Applications

Hirotaka Sakaue

Additional information is available at the end of the chapter

http://dx.doi.org/10.5772/57416

1. Introduction

In aerospace engineering, anodized-aluminum pressure-sensitive paint (AA-PSP) has been used in short duration time tests [1 - 12], unsteady flow visualizations, and unsteady pressure measurements [13 – 25]. Because of its nano-open structure (Figure 1), AA-PSP yields high mass diffusion that results in a pressure response time on the order of ten microseconds [26]. This structure enables oxygen gas to interact directly with luminophores on the pore surface, which provides fast response to pressures. By applying an AA-PSP, we can obtain global surface pressure information instead of pointwise information that may result in wide applications in pressure detection fields. AA-PSP is an optical sensor that consists of a molecular pressure probe (luminophore) and an anodized aluminum as a supporting matrix. As schematically shown in Figure 2, the luminophore on the anodized-aluminum surface is excited by an illumination source and gives off luminescence. This luminescence is related to gaseous oxygen in a test gas, a process called oxygen quenching. Because the gaseous oxygen can be described as a partial pressure of oxygen as well as a static pressure, the luminescence from an AA-PSP can be described as a static pressure. See Section 3.2 for a detailed description.

The luminophore is directly related to important parameters of AA-PSPs, such as the luminescent signal level, pressure sensitivity, temperature dependency, and response time. Mainly three types of luminophores are commonly used for PSP in general, such as ruthenium complex, porphyrin, and pyrene. Each luminophore has an optimum excitation wavelength, and its peak wavelength of luminescence varies by the luminophore as well. For AA-PSP, the luminophore is applied on the anodized-aluminum surface by the dipping deposition method [27]. This method requires a luminophore, a solvent, and an anodized-aluminum coating. The

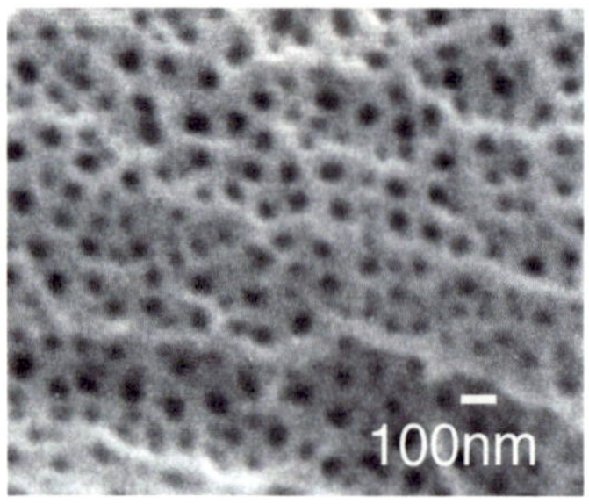

Figure 1. Nano-open structure of anodized-aluminum surface. Surface image was taken using a scanning electron microscope.

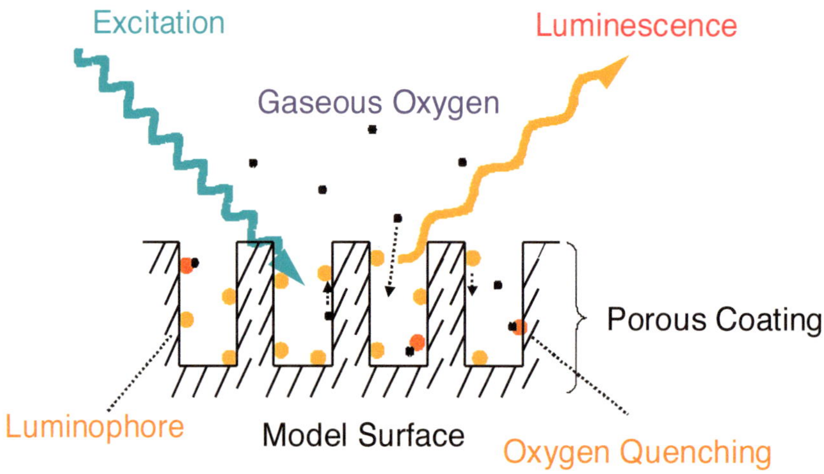

Figure 2. Schematic description of anodized-aluminum pressure-sensitive paint (AA-PSP).

application procedure is schematically shown in Figure 3. This method first dissolves the luminophore in solvent, and anodized-aluminum coating is dipped in the solution to apply the luminophore on the anodized-aluminum surface. However, this method was not quite understood, so that other luminophores were not successfully applied on an anodized-aluminum surface. For a given luminophore, a selection of solvent may influence to the AA-PSP characterizations: the signal level, pressure sensitivity, temperature dependency, and response time. The luminophore concentration may influence to the AA-PSP characterizations, because the amount of luminophore on an anodized-aluminum surface may change with the concentration used in the dipping deposition. The dipping duration can be another important parameter that influences the AA-PSP characterizations, because it would influence the amount of luminophore applied on the anodized-aluminum surface. The effects on the dipping duration as well as the above mentioned dipping parameters would give us fundamental knowledge to apply various luminophores on the anodized-aluminum coating. However, the effects of these parameters on AA-PSP have not been studied.

http://dx.doi.org/10.5772/57416

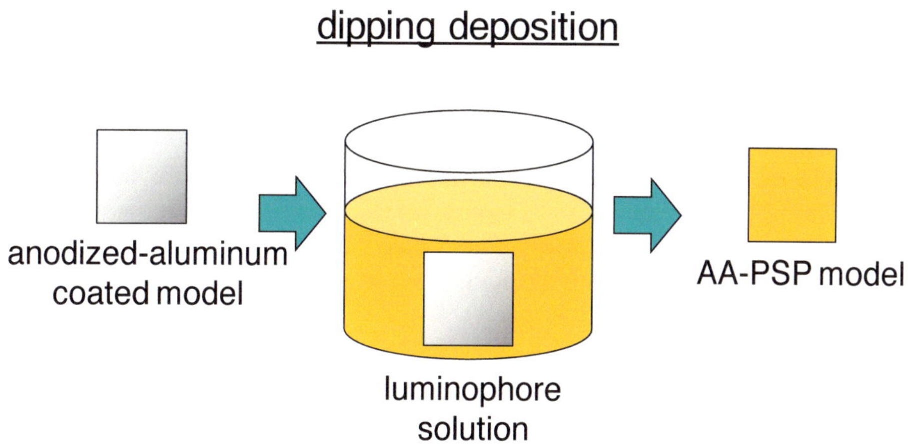

Figure 3. Schematic description of dipping deposition method.

In this chapter, the luminophore application method of dipping deposition is studied. This includes steady- and unsteady-state characterizations of AA-PSP, such as the signal level, pressure sensitivity, temperature dependency, and response time. The study of this method will expand our selection of luminophores onto an anodized aluminum, which will be beneficial for fabricating AA-PSP for various aerodynamic-measurement purposes especially in unsteady measurement applications.

2. Materials and luminophore application

Anodized-aluminum coated samples were prepared from a sheet of pure aluminum, anodization process, and cut in pieces (10 mm × 10 mm). The anodization process and the luminophore-application process were as follows. The anodization process gives an anodized aluminum of 20 nm pore size with the thickness of 10 ± 1 μm. The thickness was measured by an eddy current apparatus (Kett, LZ-330). Bathophen ruthenium (RuDPP) from GFS Chemicals was used as a luminophore. It is a conventional luminophore for AA-PSP.

- **Step1 Pretreatment.** An aluminum sheet was dipped in a 2% sodium hydroxide solution for 2 min to remove an excess oxidized layer. The surface was rinsed by water after this process.
- **Step2 Anodization.** A constant current density (10 mA/cm^2) was applied to the aluminum sheet, which was connected to anode in 1 M sulfuric acid at 0 °C.
- **Step3 Anodized layer modification.** The anodized sheet was dipped in a 5 % phosphoric acid for 20 min at 30 °C. After this process, the sheet was rinsed by water.
- **Step4 Luminophore application (dipping deposition).** For a given luminophore (RuDPP), the solvent polarity, luminophore concentration, and dipping duration were varied to study

the luminophore application process. After the dipping, AA-PSP was rinsed with the same solvent used. Then, remained solvents on AA-PSP samples were evaporated in a vacuum chamber at 50 °C for about 3 hours.

In total eight solvents (hexane, toluene, dichloromethane, chloroform, acetone, *N,N*-dimethyl formamide, dimethyl sulfoxide, and water) were selected for the solvent-polarity study in order from non-polar to the highest polarity index. The luminophore concentration was selected from a very dilute case of 0.001 mM to 10 mM, where the luminophore reached to its saturation. The dipping duration was varied from a very short dipping of 1 s to a very long dipping of 100,000 s (over 1 day). Even though the upper limit of the dipping duration would be infinity, the author assumed that over 1 day of dipping duration would be enough to understand the change in the AA-PSP characterizations. The reference AA-PSP, which was labeled as $AAPSP_{ref}$, was created by dichloromethane as a solvent, the concentration of 0.1 mM, and the dipping duration of one hour.

To study the effect on the solvent polarity, 11.7 mg of RuDPP was dissolved in 100 ml of eight different solvents based on the polarity index (Table 1 (a)). If RuDPP is dissolved completely, the concentration was 0.1 mM. If not dissolved, the solution was saturated with excess RuDPP remained. Anodized-aluminum samples were dipped in RuDPP solutions. The dipping time was one hour at room conditions. Eight different AA-PSP samples were labeled based on their polarity index of solvents (Table 1 (a)).

To study the effect on the luminophore concentration, dichloromethane was chosen as a solvent. The concentration had the range of the fifth order of magnitude; it was varied from 0.001 mM to 10 mM. The dipping duration was one hour at room conditions. Table 1 (b) lists the luminophore application conditions related to the concentration. Prepared AA-PSPs were labeled (also listed in Table 1 (b) as Sample ID).

To study the effect on the dipping duration, dichloromethane was chosen as a solvent, and the concentration of the luminophore solution was fixed at 0.1 mM. The duration was varied from 1 s to 100,000 s. Table 1 (c) lists the conditions related to the dipping duration. Prepared AA-PSPs are labeled based on their dipping conditions, which are also listed in Table 1 (c) as Sample ID.

3. Steady-state characterization

Figure 4 schematically describes the calibration system, which consists of a spectrometer (Hitachi High Technologies, F-7000) and a pressure- and temperature-controlled chamber. This system characterizes the luminescent spectrum of an AA-PSP sample with varying pressures and temperatures. For characterization, an AA-PSP sample was placed in the test chamber. The excitation wavelength was set at 460 nm by a monochromator via a xenon lamp illumination in the spectrometer unit. The chamber has optical windows that passed the excitation from the illumination unit and the luminescence from the sample. The luminescence from AA-PSP samples was measured from 570 to 800 nm for a given pressure and a given

http://dx.doi.org/10.5772/57416

Sample ID	Polarity Index	Solvent
$AAPSP_{ind00}$	0.1	Hexane
$AAPSP_{ind02}$	2.4	Toluene
$AAPSP_{ref}$	3.1	Dichloromethane
$AAPSP_{ind04}$	4.1	Chloroform
$AAPSP_{ind05}$	5.1	Acetone
$AAPSP_{ind06}$	6.4	*N*,*N*-dimethylformamide
$AAPSP_{ind07}$	7.2	Dimethylsulfoxide
$AAPSP_{ind10}$	10.2	Water

(a)

Sample ID	Luminophore Concentration (mM)
$AAPSP_{00.001}$	0.001
$AAPSP_{00.010}$	0.01
$AAPSP_{ref}$	0.1
$AAPSP_{01.000}$	1
$AAPSP_{10.000}$	10

(b)

Sample ID	Dipping Duration (s)
$AAPSP_{1}$	1
$AAPSP_{10}$	10
$AAPSP_{100}$	100
$AAPSP_{1000}$	1,000
$AAPSP_{ref}$	3,600
$AAPSP_{100000}$	100,000

(c)

Table 1. (a). Luminophore application conditions: solvent polarity. Dipping solvent was selected based on the polarity index. RuDPP concentration was fixed at 0.1 mM, and anodized-aluminum coatings were dipped at room temperature for one hour. (b). Luminophore application conditions: luminophore concentration. RuDPP concentration was varied from 0.001 mM to 10 mM. Dipping solvent was dichloromethane, and anodized-aluminum coatings were dipped at room temperature for one hour.(c). Luminophore application conditions: dipping duration. Dipping duration was varied from 1 to 100,000 s. Dichloromethane was chosen as a solvent, and RuDPP concentration was fixed at 0.1 mM.

temperature. The luminescent signal of an AA-PSP was then determined by integrating the spectrum from 600 to 700 nm. For pressure calibration, the chamber was connected to a pressure controlling unit (Druck DPI515), with settings from 5 to 120 kPa at a constant temperature at 25 °C. For temperature calibration, a sample heater/cooler was controlled to vary the temperature from 10 to 50 °C with a constant pressure at 100 kPa. The test gas was dry air. For the signal level characterization, all the AA-PSP samples were measured with the same optical setup in the spectrometer but replacing samples in the chamber at constant pressure and temperature of 100 kPa and 25 °C, respectively. Throughout our characterizations, reference conditions were 100 kPa and 25 °C. The signal level, η, pressure sensitivity, σ, and temperature dependency, δ, were characterized from the luminescent signals of AA-PSPs. Definitions and procedures to derive these characterizations are described in Sections 3.1, 3.2, and 3.3.

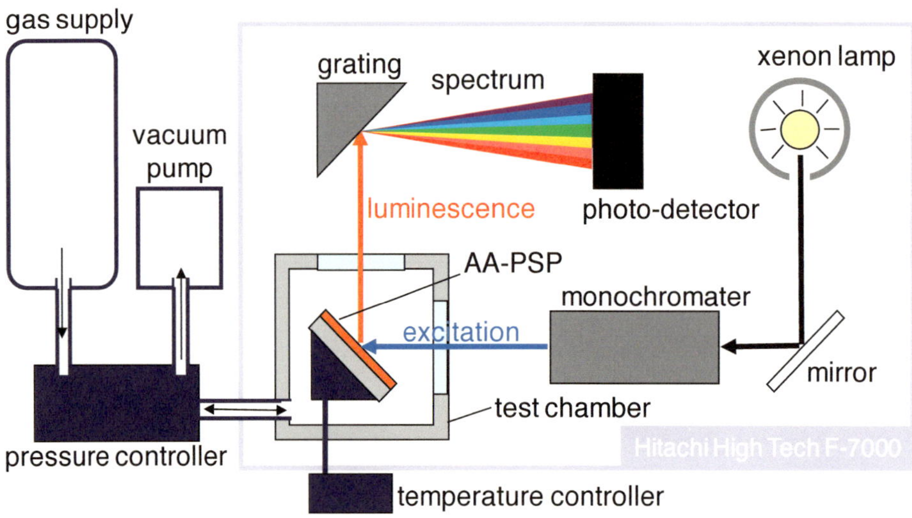

Figure 4. Schematic of AA-PSP calibration setup.

3.1. Signal Level

The luminescent signal, I, was determined by the integration of AA-PSP spectrum from 600 to 700 nm. Based on Liu *et al.*, this can be described by the gain of the photo-detector in our spectrometer, G, the emission from AA-PSP, I_{AAPSP}, the excitation in the spectrometer, I_{ex}, and the measurement setup component, f_{set} [28]:

$$I = G I_{AAPSP} I_{ex} f_{set} \tag{1}$$

In our calibration setup, G, I_{ex}, and f_{set} were the same for all AA-PSP samples. We non-dimensionalized the luminescent signal by that of AAPSP$_{\mathrm{ref}}$, $I_{AAPSPref}$. All luminescent signals were determined at the reference conditions. We call this value as the signal level, η, shown in Equation (2):

$$\eta = \frac{I}{I_{AAPS\,\mathrm{Pref}}} \ (\%) \tag{2}$$

http://dx.doi.org/10.5772/57416

3.2. Pressure sensitivity

Based on the Stern-Volmer relationship, the luminescent intensity, *I*, is related to a quencher [29]:

$$\frac{I_0}{I} = 1 + Kq[O2] \tag{3}$$

Where I_0 is the luminescent intensity without quencher and K_q is the Stern-Volmer quenching constant. The quencher is oxygen, which is described by the oxygen concentration, $[O_2]$. For AA-PSP, $[O_2]$ can be described by the adsorption and surface diffusion of the adsorbed oxygen on an anodized-aluminum surface. We can describe $[O_2]$ by the partial pressures of oxygen as well as the static pressures. These are combined with Equation (3) to give the adsorption-controlled model [27]:

$$\frac{Iref}{I} = A + B\left(\frac{p}{p_{ref}}\right)^{\gamma} \tag{4}$$

Where *A*, *B*, and γ are calibration constants, respectively. Here, *ref* denotes our reference conditions.

Pressure sensitivity, σ (%), describes the change in the luminescent signal over a given pressure change. This corresponds to a slope of the Equation (4) at the reference conditions:

$$\sigma = \left.\frac{d(I_{ref}/I)}{d(p/p_{ref})}\right|_{p=p_{ref}} = B\cdot\gamma \quad (\%) \tag{5}$$

3.3. Temperature dependency

AA-PSP, like PSP in general, has a temperature dependency [30]. This influences the luminescent signal, which can be described as the third order polynomial in Equation (6):

$$\frac{I}{I_{ref}} = c_{T0} + c_{T1}T + c_{T2}T^2 + c_{T3}T^3 \tag{6}$$

Where c_{T0}, c_{T1}, c_{T2}, and c_{T3} are calibration constants, respectively. We defined the temperature dependency, δ, which is a slope of the temperature calibration at the reference conditions (Equation (7)). If the absolute value of δ is large, it tells us that the change in luminescent signal over a given temperature change is also large. This is unfavorable condition as a pressure sensor. On the contrary, zero δ means that AA-PSP is not temperature dependent:

$$\delta = \left.\frac{d(I/I_{ref})}{dT}\right|_{T=T_{ref}} = c_{T1} + 2c_{T2}T_{ref} + 3c_{T3}T_{ref}^2 \ (\%/{}^\circ C) \tag{7}$$

Overall, our δs showed negative (see Section 5.3). This means that δ_{min} is the most temperature dependent and δ_{max} the least temperature dependent.

4. Unsteady-state characterization

A vertical shock tube for characterizing the response time is schematically shown in Figure. 5 (a). The length of the driver and driven sections are 1420 mm and 5530 mm, respectively. The driven section has a square cross section of 100 mm × 100 mm, and a test section is installed at the end of the driven section. The test gas was dry air and initially set at room conditions. When the diaphragm between the driver and the driven sections is ruptured, a planar shock wave propagates into the driven section. We set the driver pressure as 400 kPa that created a planar shock wave with the Mach number of 1.30.

The schematic description of the test section is shown in Figure. 5 (b). An AA-PSP sample was fixed on a flat plate placed on the bottom wall of the shock tube. The samples were illuminated by a continuous 400 nm laser. A planar shock wave and its normal reflection created a step change of pressure. A photomultiplier tube (PMT, Hamamatsu R7236) was used to detect the intensity change of luminescence from the AA-PSP sample through a 605 ± 40 nm band-pass filter. The output signal from the PMT was amplified by Hamamatsu C1053-03 through an analog low-pass filter with a cutoff frequency of 1 MHz. The filtered signal was then digitized to 12 bits and sampled on an A/D converter (Yokogawa, DL1540C) at a rate of 200 MHz [31].

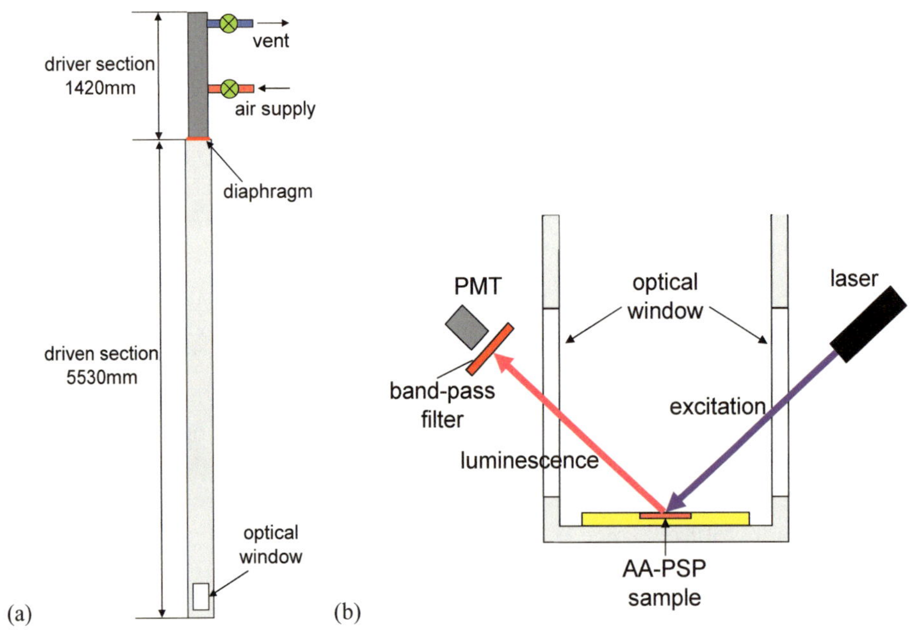

Figure 5. (a). Schematic description of shock tube. (b). Test section and optical setup of the shock tube.

http://dx.doi.org/10.5772/57416

4.1. Response time

The luminescent signal was converted to a normalized pressure, p_{norm} to characterize the response time, derived from Equation (8).

$$p_{norm} = \frac{p - p_{\min}}{p_{\max} - p_{\min}} = \frac{(I_{ref} / I - A)^{1/\gamma} - (I_{ref} / I_{\min} - A)^{1/\gamma}}{(I_{ref} / I_{\max} - A)^{1/\gamma} - (I_{ref} / I_{\min} - A)^{1/\gamma}} \tag{8}$$

Where *min* and *max* denote the minimum and maximum values of a step change, respectively. We used the 90 % rise of p_{norm} to determine the response time.

5. Characterization results

Photographs of RuDPP solution were taken to qualitatively verify the solubility of RuDPP (Figure 6). All solvents dissolved RuDPP except for the solvent with lowest polarity index (hexane). Toluene, which was the second lowest solvent in our test, partially dissolved RuDPP. Water, which gave the highest polarity index in our test, partially dissolved RuDPP, but it dissolved RuDPP completely after about one day. Other solvents dissolved RuDPP as soon as RuDPP particles were dropped to the solvents.

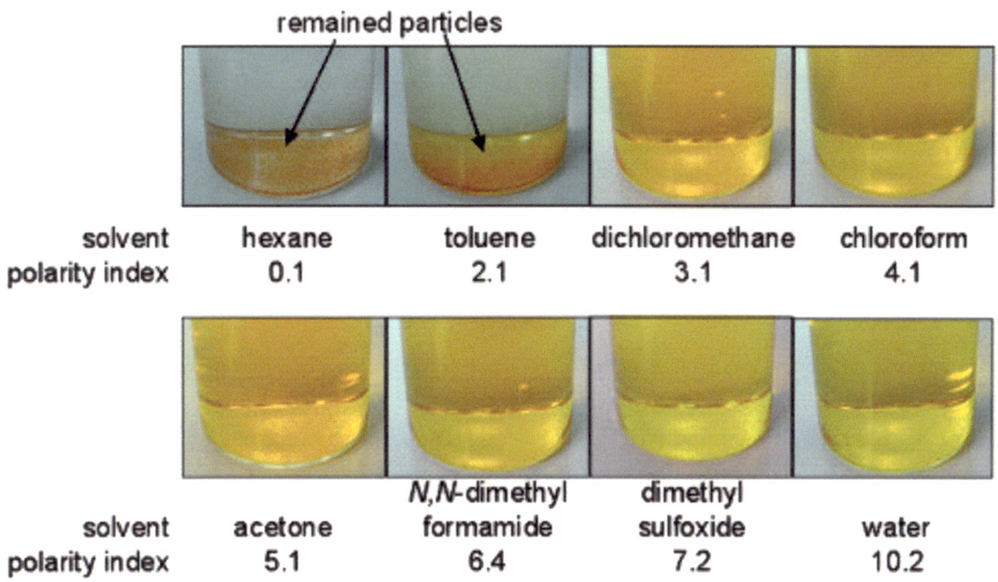

Figure 6. RuDPP dissolved in solvents with range of polarity index.

Figure 7 (a) and (b) show luminescent spectra of $AAPSP_{ref}$ with varying pressures and temperatures, respectively. Spectra were normalized by the luminescent peak at the reference conditions. We can see that, as increasing the pressure, the luminescent spectrum decreased due to oxygen quenching [29]. As the temperature increases, we can see the spectrum decreased due to the thermal quenching [29]. It was noticed that the luminescent peak was shifted

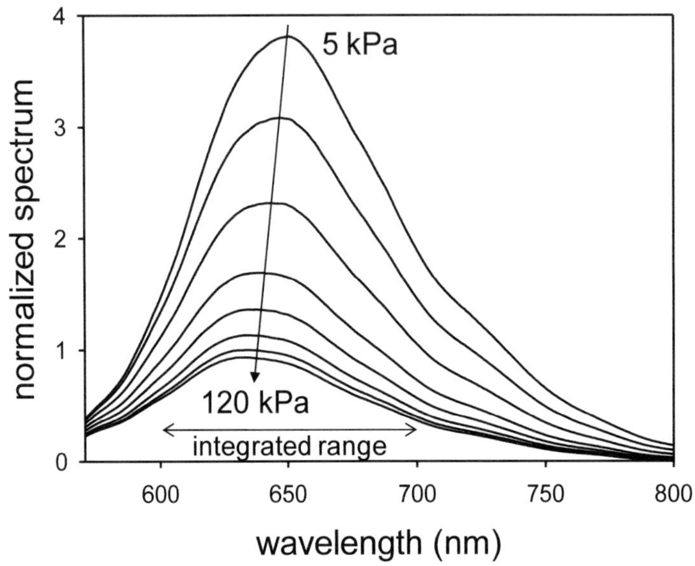

(a)

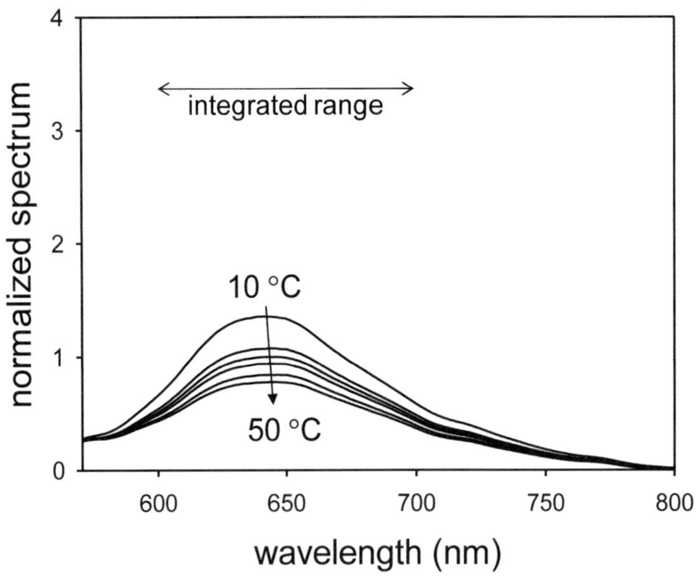

(b)

Figure 7. (a). Pressure spectra of $AAPSP_{ref}$. Thick line shows the spectrum at reference conditions of 100 kPa and 25 °C. (b). temperature spectra of $AAPSP_{ref}$. Thick line shows the spectrum at reference conditions of 100 kPa and 25 °C.

from 650 to 635 nm by increasing the pressure from 5 to 120 kPa. For temperature spectra, the peak was shifted from 640 to 645 nm by increasing the temperature from 10 to 50 °C. As described in Section 3, we integrated an obtained spectrum from 600 to 700 nm to determine as the luminescent intensity, I, for a given pressure and a temperature.

5.1. Luminescent signal

The signal level, η, was determined from Equation (2). Figure 8 (a) shows the signal level, η, related to the polarity index, normalized by the signal of $AAPSP_{ref}$ at 100 kPa. The η was shown as a bar with the determined value. It is obvious that the luminophore application method of dipping deposition greatly influenced the signal level. $AAPSP_{ind00}$ from the polarity index of 0.1 (hexane) showed very low RuDPP application on anodized aluminum, indicated by the signal level. Note that hexane did not dissolve RuDPP (Figure 6). $AAPSP_{ind02}$ from the polarity index of 2.4 (toluene) applied RuDPP well on anodized aluminum, which can be seen from the signal level. Here, toluene partially dissolved RuDPP. As increasing polarity index, RuDPP applied on anodized aluminum. However, the application suddenly dropped between polarity index at 6.4 of *N, N*-dimethyl amide and at 7.2 of dimethyl sulfoxide. The application brought back at the highest polarity index of 10.2 (water). The highest signal level of 1.89 was obtained from $AA\text{-}PSP_{ind05}$. It can be said that a range of polarity index of solvent exists that applies RuDPP. However, this range does not correspond to the range of dissolving RuDPP.

It is assumed that RuDPP remains as solution if it is dissolved well in a solvent. This assumption can be supported that *N, N*-dimethyl amide and dimethyl sulfoxide did not apply RuDPP well onto anodized aluminum. On the other hand, RuDPP applies onto anodized aluminum if it is partially dissolved in a solvent. This can be supported that toluene and water applied RuDPP onto anodized aluminum, even though these dissolved RuDPP partially. If a solvent did not dissolve RuDPP, it would not be applied onto anodized aluminum. This can be seen from the result of hexane.

As we increased the luminophore concentration from 0.001 mM to 0.1 mM, η increased (Figure 8 (b)). Note that the vertical axis in Figure 8 (b) was shown as log scale. The η was shown as a bar with the determined value. Even though we increased the concentration more than 0.1 mM, η decreased roughly by a half. This may be due to the concentration quenching [29]. There was an optimum concentration to maximize η. The maximum η was obtained from $AAPSP_{00.100}$, whose luminophore concentration was 0.1 mM.

There was a peak dipping duration to maximize η (Figure 8 (c)). Note that the vertical axis in Figure 8 (c) was shown as log scale. The η was shown as a bar with the determined value. The maximum η was obtained from $AAPSP_{1000}$, whose dipping duration was 1,000 s. For a short dipping duration, the luminophore would remain in the luminophore solution instead of applying onto the anodized surface. Roughly, the difference of η was a factor of 8.5 by varying the dipping duration. Even though we increased the dipping duration over 1,000 s, η decreased. This may be due to the concentration quenching, influencing to the luminophore application [29].

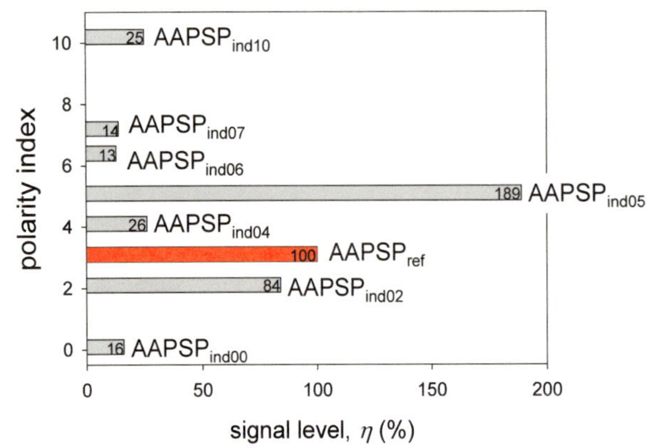

(a)

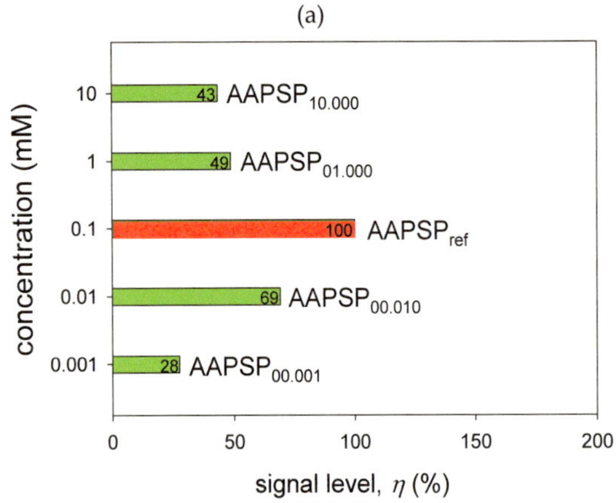

(b)

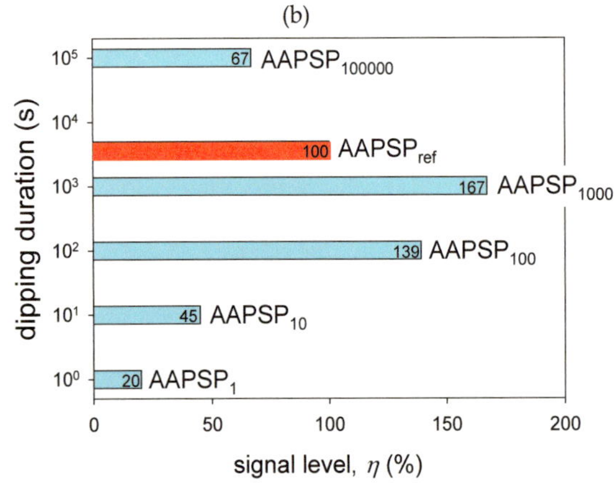

(c)

Figure 8. (a). Signal level, η, related to the polarity index. (b). Signal level, η, related to the luminophore concentration. (c). Signal level, η, related to the dipping duration

http://dx.doi.org/10.5772/57416

5.2. Pressure calibration

Figure 9 shows pressure calibrations related to the polarity index, fitted with the adsorption controlled model in Equation (4). The reference was set at atmospheric conditions. The relationship between the luminescent ratio, I_{ref}/I, and the pressure ratio, p/p_{ref}, was non-linear at low pressure region. We can see that the calibration was influenced by the solvent polarity.

The pressure sensitivity, σ, was shown as a bar with the determined value from Equation (5) (Figure 9). The solvent polarity greatly influenced σ, even though the same luminophore was applied onto the same anodized aluminum. The highest σ of 0.62 was obtained from $AAPSP_{ref}$. This showed the peak sensitivity as varying the solvent polarity. Another peak was seen at the polarity index of 10.2 (water).

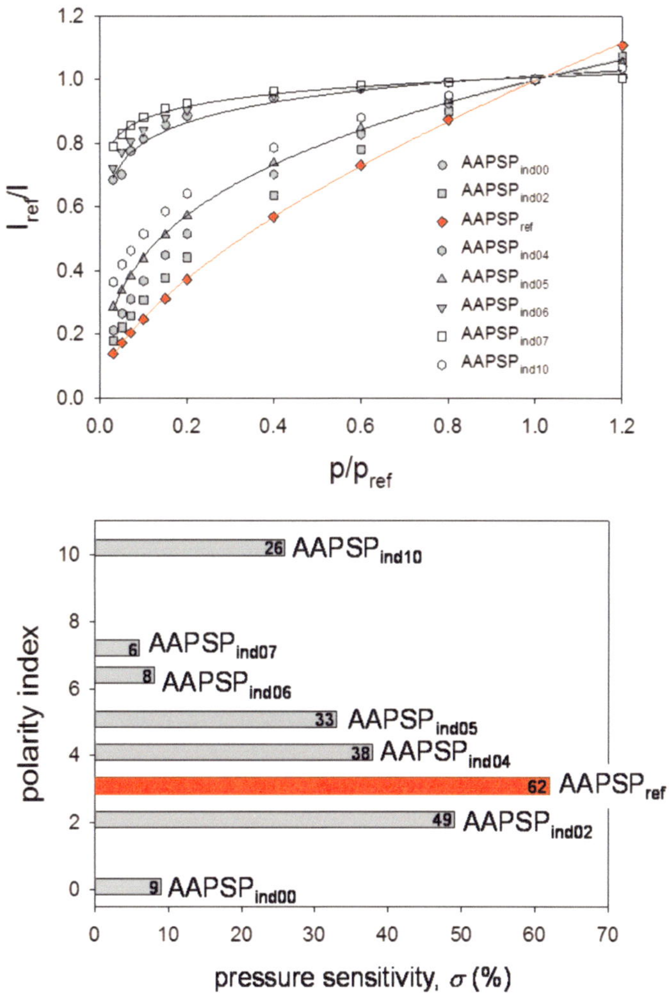

Figure 9. Pressure calibration and pressure sensitivity, σ, related to the polarity of solvent.

Figure 10 shows pressure calibrations related to the luminophore concentration, fitted with the adsorption controlled model in Equation (4). We can see two groups in calibrations: the luminophore concentration up to 0.1 mM and the concentration higher than 0.1 mM. The former showed steeper calibrations than the latter. This tells us that the former group was more pressure sensitive than the latter.

The pressure sensitivity, σ, was determined by using Equation (5). This value was listed in the bar scale (Figure 10). AA-PSP with the luminophore concentration up to 0.1 mM showed σ around 60%, while AA-PSP with higher concentration than 0.1 mM showed σ around 30%. This tells us that even though the amount of luminophore over 0.1 mM was dissolved in the dipping solution, σ did not increase. The decrease in σ may be due to the concentration quenching [29].

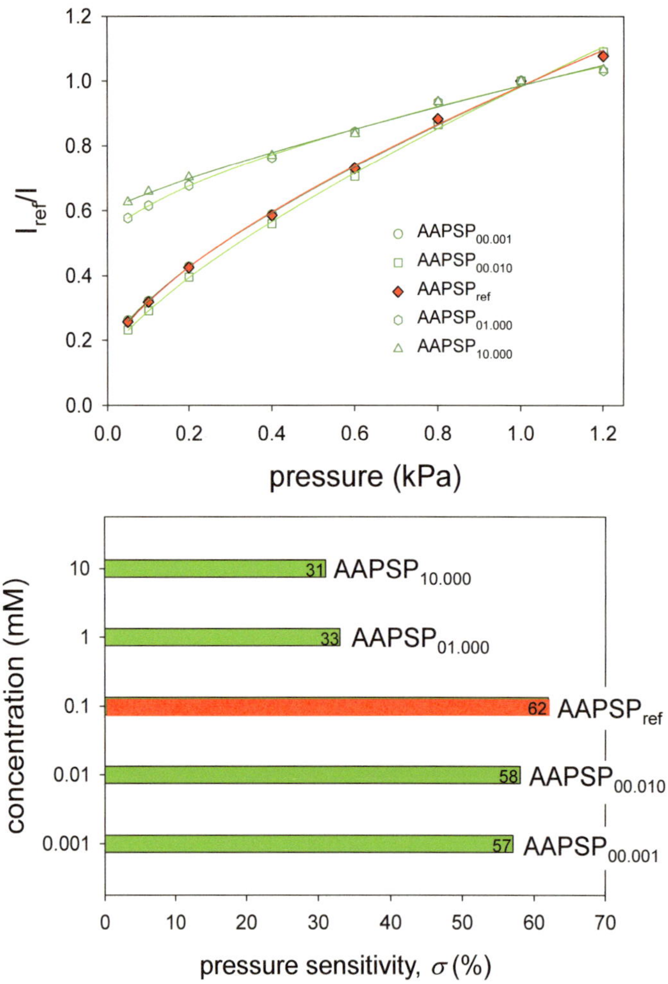

Figure 10. Pressure calibration and pressure sensitivity, σ, related to the he luminophore concentration.

http://dx.doi.org/10.5772/57416

Figure 11 shows the pressure calibrations related to the dipping duration. The value of σ was determined from Equation (5). This value was shown as a bar scale in Figure 11. The maximum σ of 65% and the minimum σ of 52% were obtained from $AAPSP_{100}$ and $AAPSP_1$, respectively. Even though the fifth order difference in the dipping duration was provided, a minimal effect was seen on the pressure sensitivity.

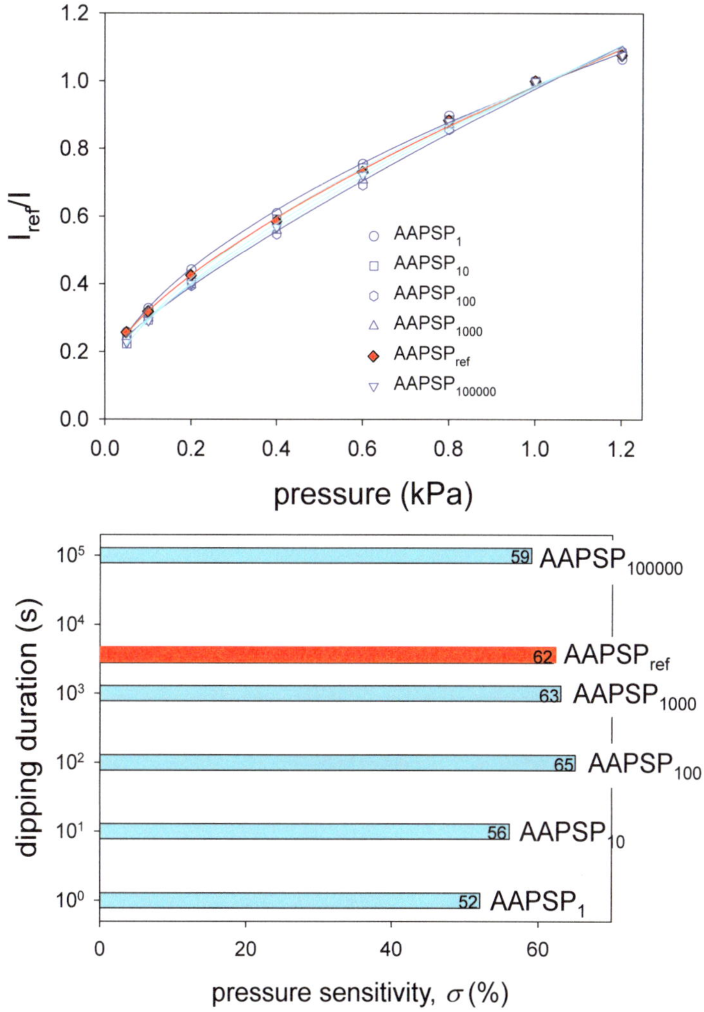

Figure 11. Pressure calibration and pressure sensitivity, σ, related to the dipping duration.

5.3. Temperature calibration

Figure 12 shows temperature calibrations related to the polarity index. Calibration plots were fitted with the third order polynomial described in Equation (6). We can see a monotonic decrease of the luminescent signal with increase of the temperature. The temperature dependency, δ, was determined from Equation (7), which was listed as a bar scale in Figure 12. The δ showed a similar tendency to σ by varying the solvent polarity. This tells us that AA-PSP is more temperature dependent if it is more pressure sensitive.

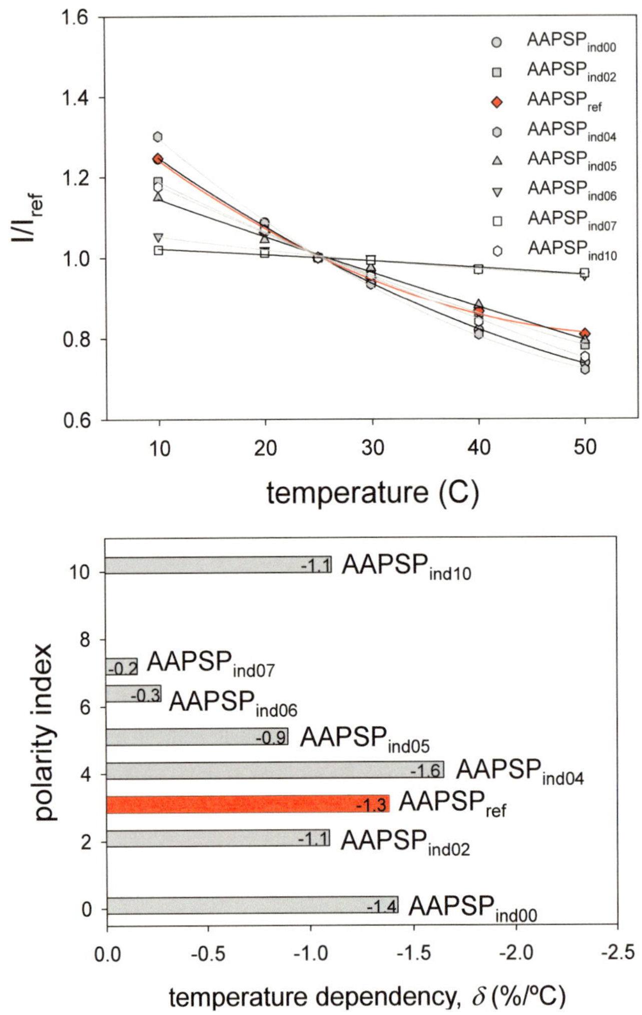

Figure 12. Temperature calibration and temperature dependency related to the polarity index.

Figure 13 shows temperature calibrations related to the luminophore concentration. Calibration plots were fitted with Equation (6). The calibrations show a monotonic decrease in luminescent signal as the temperature increased. As the concentration decreases, the calibrations became steep. This tells us that the temperature dependency tends to increase as the luminophore concentration decreases.

The temperature dependency, δ, was determined from Equation (7), which was listed as a bar scale in Figure 13. As we increased the luminophore concentration, δ decreased. Roughly, δ became more than a half by setting the luminophore concentration from 0.001 to 10 mM.

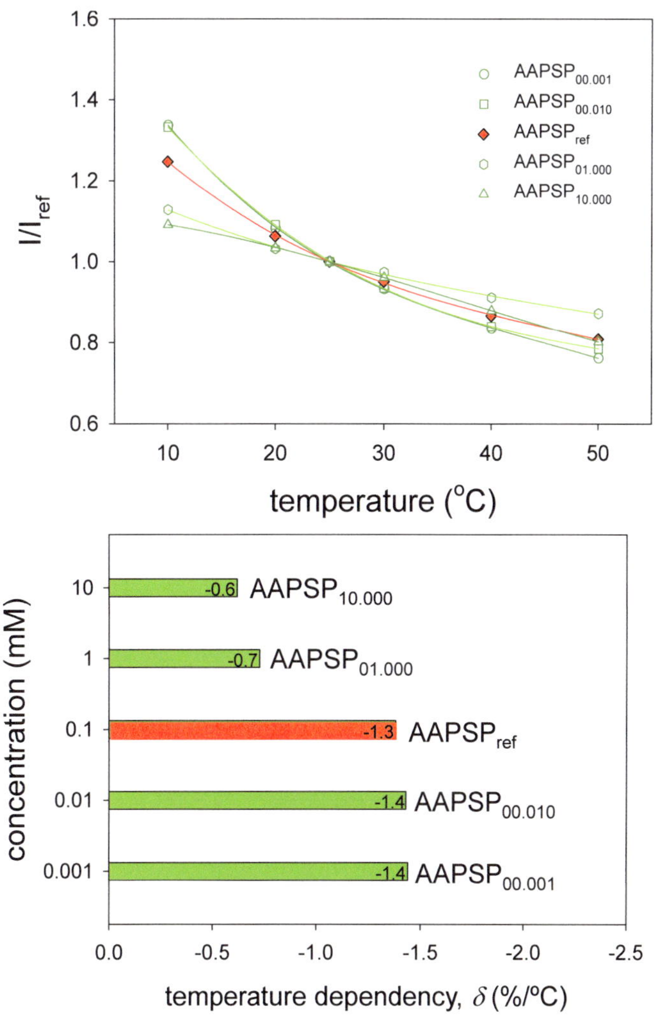

Figure 13. Temperature calibration and temperature dependency related to the luminophore concentration.

Figure 14 shows the temperature calibrations related to the dipping duration. The calibrations were fitted with Equation (6). The temperature calibrations showed the decrease in I with increase of the temperature.

The value of δ was determined from Equation (7), which was listed as a bar scale in Figure 14. With increase the dipping duration, we can see that δ decreased until 100 s and increased over this dipping duration. The difference of δ was roughly a factor of 2. Compared to the effect on the pressure sensitivity, the dipping duration showed a greater effect on the temperature dependency.

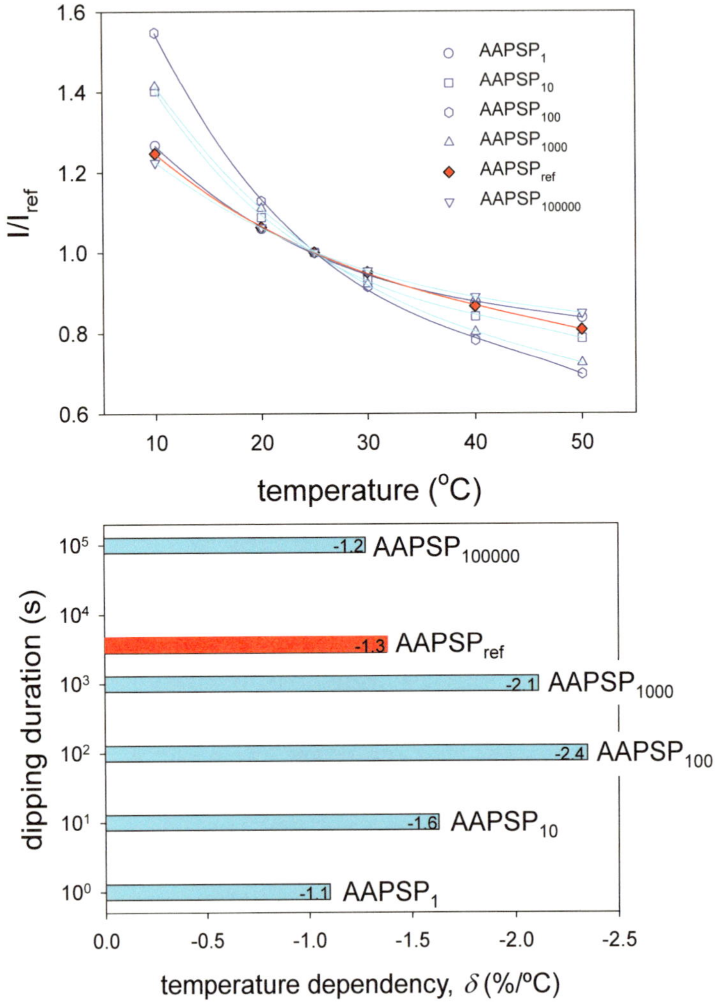

Figure 14. Temperature calibration and temperature dependency related to the dipping duration.

5.4. Response time

A normalized step change of pressure, p_{norm}, was converted from luminescent signals through the Equation (8). Figure 15 shows response time results of AA-PSPs. Because each shock tube measurement was done by a single measurement, electrical noise with high frequency content still existed. This limited the response time characterization. The limited results were shown from the solvent polarity: AAPSP$_{ind02}$, AAPSP$_{ind03}$, AAPSP$_{ind04}$, and AAPSP$_{ind05}$. These had relatively high signal levels and pressure sensitivities to separate their luminescent signal

http://dx.doi.org/10.5772/57416

changes from the electrical noise. Response times of AA-PSP were determined at the 90 % rise of p_{norm} (Section 4.1). The results ranged from 30 to 40 μs. In our setup, the thickness of anodized aluminum was 10 μm. This had ±10 % uncertainty from our instrument (Kett LZ-330). Kameda *et al.* reported that response time of AA-PSP is proportional to the squared value of its thickness, which corresponds to about ±20 % uncertainty in our measurement results [26]. Response time results were within this uncertainty, even though there were variations in step response of AA-PSPs. Considering the thickness uncertainty, response times of AA-PSPs can be said on the order of ten microseconds with the anodized-aluminum thickness of 10 μm. Sakaue *et al.* reported that the response time of AA-PSP showed a minimal effect by the luminophore selected [31]. This indicates that the response time of AA-PSP has smaller effect on the luminophore application parameters than the effect on AA-PSP thickness.

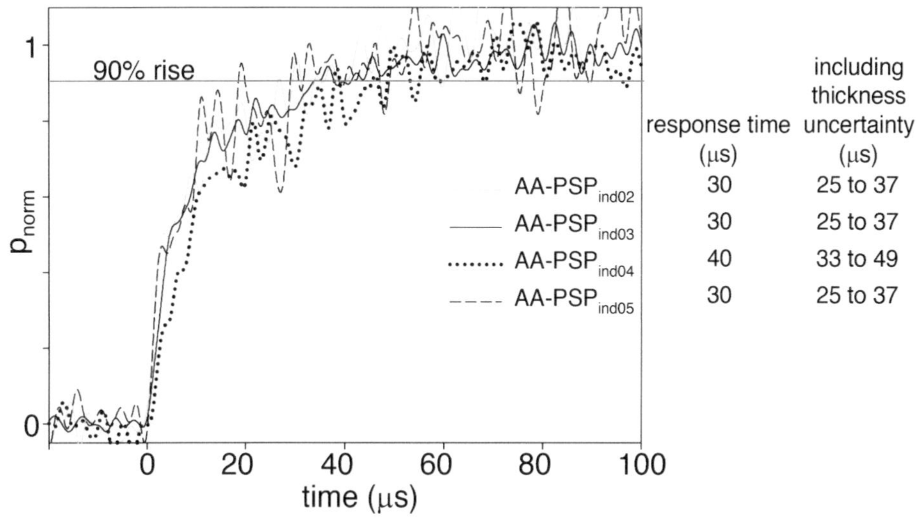

Figure 15. Response time results related to the solvent polarity.

6. Discussion: AA-PSP characterizations related to Luminophore Application Parameters

Table 2 lists the maximum and minimum values of AA-PSP characterizations: signal level, pressure sensitivity, and temperature dependency. The signal level was greatly influenced by varying the solvent polarity. The difference was a factor of 14.5. The second largest effect was the dipping duration. The difference in the signal level was a factor of 8.4. The luminophore concentration influenced the signal level for a factor of 3.6. Overall, the signal level was most influenced by the luminophore-application parameters.

The pressure sensitivity was greatly influenced by the solvent polarity. The difference in the sensitivity was a factor of 10.3. By varying the luminophore concentration, the difference was a factor of 2. The difference was a factor of 1.2 by varying the dipping duration. Among the dipping parameters, the pressure sensitivity was most influenced by the solvent polarity.

The temperature dependency was greatly influenced by the solvent polarity. The difference was a factor of 8. The differences by the luminophore concentration and the dipping duration were on the same order, which was a factor of 2. The temperature dependency was influenced by the dipping parameters, but the change was not as large as that of the pressure sensitivity. Overall, the solvent polarity influenced the most of the AA-PSP characterizations.

	Solvent Polarity		**Luminophore Concentration**		**Dipping Duration**	
	max.	**min.**	**max.**	**min.**	**max.**	**min.**
η (%)	189	13	100	27.5	167.1	20.1
σ (%)	62	6	62	31	65	52
δ (%/°C)	-0.2	-1.6	−0.6	−1.4	−1.1	−2.4

Table 2. The maximum and minimum AA-PSP characterizations.

7. Conclusions

The luminophore application method of dipping deposition was studied to provide the relationship between this method and AA-PSP characterizations for fabricating an optimized optical pressure sensor for unsteady aerodynamic applications. The characterizations were the signal level, pressure sensitivity, temperature dependency, and response time. Three important parameters in the luminophore application method were studied: solvent polarity, luminophore concentration, and dipping duration. It was found that the AA-PSP characterizations were related to one another. Therefore, an absolute optimization of the luminophore application method was not obtained. However, the relationship among these characterizations and the luminophore-application parameters were revealed, which were concluded as follows.

The solvent polarity was the most influencing parameter. The signal level showed the widest range from 13% to 189% compared to the signal level of the reference AA-PSP (100%). The pressure sensitivity ranged from 6 to 62 %, and the temperature dependency from -0.2 to -1.6 %/°C. It was seen that the pressure-sensitive AA-PSP was also temperature sensitive. It was shown qualitatively by photograph that the solubility was related to the solvent polarity. Well luminophore-dissolved solvents did not show higher AA-PSP outputs. This may be that the luminophore remained in the solvent and was not applied onto the anodized-aluminum surface well.

The luminophore concentration and dipping duration greatly influenced to the signal level. However, the influence to the pressure sensitivity and the temperature dependency was relatively small. The difference was less than or equal to a factor of 2.

The effect of AA-PSP response time due to the dipping deposition method was smaller than the effect by the thickness uncertainty of AA-PSP. With the anodized-aluminum thickness of 10 μm, the response time characterization was within the thickness uncertainty. The response time was on the order of ten microseconds.

Acknowledgements

The author would like to thank his colleagues for technical supports: Dr. K. Morita (JAXA), Mr. Y. Iijima (JAXA), Ms. K. Ishii (The University of Tokyo), Mr. Y. Yamada (The University of Electro-Communications), and Prof. Y. Sakamura (Toyama Prefectural University).

Author details

Hirotaka Sakaue*

Address all correspondence to: sakaue@chofu.jaxa.jp

Institute of Aeronautical Technology, Japan Aerospace Exploration Agency / Chofu, Tokyo, Japan

References

[1] Nakakita K, *et al*. 2000. AIAA2000-2523.

[2] Nakakita K, Asai K. 2002. AIAA2002-2911.

[3] Ishiguro Y, *et al*. 2007. AIAA2007-01187.

[4] Miyamoto K, *et al*. 2010. AIAA2010-4798.

[5] Morita K, *et al*. 2011. AIAA2011-3724.

[6] Disotell KJ, Gregory JW. 2011. Rev. Sci. Instrum. 82:075112.

[7] Disotell KJ, *et al*. 2012. AIAA2012-2757.

[8] Yang L, *et al*. 2012. Int. J. Heat Fluid Flow 37: 9 – 21.

[9] Yang L, *et al*. 2012. Sens. Actuators B 161:100 – 7.

[10] Yang L, *et al*. 2012. Exp. Therm. Fluid Sci. 40:50 –56.

[11] Hayashi T, *et al*. 2012. 28th International Symposium on Shock Waves, Vol. 1, ISBN 978-3-642-25687-5, pp. 607 – 613.

[12] Fujii S, *et al.* 2013. AIAA2013-0485.

[13] McGraw CM, *et al.* 2003. Rev. Sci. Instrum. 74:5260-66.

[14] Virgin CA, *et al.* 2005. Proc. 2005 ASME Int. Mech. Eng. Congr. Expo., pp. 297-307.

[15] McGraw CM, *et al.* 2006. Exp. Fluids 40:203-11.

[16] Nakakita K. 2007. AIAA2007-3819.

[17] Nakakita K, Arizono H. 2009. AIAA2009-3847.

[18] Klein C, *et al.* 2010. Numerical and Experimental Fluid Mechanics VII, pp. 323-30.

[19] Yorita D, *et al.* 2010. AIAA2010-0307.

[20] Asai K, Yorita D. 2011. AIAA2011-0847.

[21] Nakakita K, *et al.* 2012. AIAA2012-2758.

[22] Steimle PC, *et al.* 2012. AIAA J. 50:399-415.

[23] Watkins AN, *et al.* 2012. AIAA2012-2756.

[24] Wong OD, *et al.* 2012. Proc. 68th Am. Helicopter Soc. Annu. Forum Technol. Disp., Pap. AHS2012-000233.

[25] Mérienne MC, *et al.* 2013. AIAA2013-1136.

[26] Kameda, M.; Tezuka, N.; Hangai, T.; Asai, K.; Nakakita, K.; Amao, M. *Meas. Sci. Technol.* 2004, *15*, 489–500.

[27] Sakaue, H. *Rev. Sci. Instrum.* 2005, *76*, 084101.

[28] Liu, T.; Guille, M.; Sullivan, J.P. Accuracy of Pressure Sensitive Paint. *AIAA J.* 2001 *40*, 103–112.

[29] Lakowicz, J.R. *Principles of Fluorescence Spectroscopy.* Kluwer Academic/Plenum Publishers: New York, NY, USA, 1999; Chapter 1.4.A.

[30] Liu, T.; Sullivan, J.P. *Pressure and Temperature Sensitive Paints;* Springer Verlag: Heidelberg, Germany, 2004; pp. 27–31 and Chapter 7.

[31] Sakaue, H., Morita, K., Iijima, Y., Sakamura, Y., Sensors and Actuators A: Physical, Elsevier, Vol. 199, No. 1, pp. 74 – 79, 2013.